Empathy in Algorithms

Dimitrios Sargiotis

Empathy in Algorithms

Shaping Ethical AI for a Just and Inclusive Society

Springer

Dimitrios Sargiotis
School of Civil Engineering
National Technical University of Athens
Zografou Campus-Athens, Greece

ISBN 978-3-032-32127-5 ISBN 978-3-032-32128-2 (eBook)
https://doi.org/10.1007/978-3-032-32128-2

This Springer imprint is published by the registered company Springer Nature Switzerland AG
The registered company address is: Gewerbestrasse 11, 6330 Cham, Switzerland

Preface

In the long arc of human history, technology has always reshaped the architecture of society, recalibrating the ways we govern, work, and live together. Yet no prior invention whether the plough, the printing press, or the microprocessor has carried with it the profound ethical weight of artificial intelligence. AI is not merely an instrument of efficiency; it is a force of moral consequence. Its decisions influence who receives opportunity, who is left behind, and how the fragile covenant between citizen and state is re-imagined in the twenty-first century.

This volume, *Empathy in Algorithms: Shaping Ethical AI for a Just and Inclusive Society*, situates itself at the intersection of political philosophy, technological innovation, and civic responsibility. It asks not only what AI can do but what AI ought to do a distinction that marks the boundary between technocracy and justice, between computation and conscience.

The chapters that follow explore the multiple dimensions of this ethical terrain. They begin with the reconfiguration of public policy in an era where algorithms promise to deliver equitable and efficient services, yet simultaneously risk exacerbating structural inequalities and technological unemployment. They advance toward the delicate terrain of data governance, where encryption, anonymization, and adversarial training emerge as technical guardians of civic trust. They interrogate social inclusion and the digital divide, illuminating how the benefits of AI and smart infrastructures must be accessible to all, irrespective of socioeconomic or geographic positioning. And they culminate in frameworks for ethical AI development that foreground social impact, transparency, and regulatory foresight, insisting that technology remain aligned with human dignity.

What unites these themes is a conviction that justice cannot be abstracted away from design. To embed fairness, accountability, and empathy into algorithms is not a matter of technical refinement alone it is a philosophical act, a political commitment, and a profoundly human responsibility. By exposing the sources of bias in AI, by demanding human-centered architectures, and by engaging with inclusive governance structures, the book insists on AI not as an opaque force of inevitability, but as a deliberative instrument of collective good.

For scholars, policymakers, engineers, and citizens alike, this book offers more than a catalogue of challenges and solutions. It offers an ethical lens: a way of perceiving AI not as a neutral system of prediction, but as a moral agent embedded in civic life. It seeks to inspire a generation of researchers and practitioners to transcend the reductionism of performance metrics and to embrace the deeper measure of success whether AI helps us build societies that are not only smarter but also kinder, fairer, and more humane.

If history is to judge this century not merely by its innovations but by its compassion, then the ethical shaping of artificial intelligence will be among its most enduring tests. This volume is offered as both compass and companion in that urgent journey.

Zografou Campus-Athens, Greece Dimitrios Sargiotis

Acknowledgments Although this book bears my solitary authorship, it is in no sense the product of isolation. It emerges from a long continuum of conversations, reflections, and intellectual inheritances that extend far beyond these pages. At its heart, this work is less an individual statement than a convergence of ethical concern, philosophical inquiry, and technical imagination all directed toward one urgent question: how to shape artificial intelligence into a civic force that dignifies, rather than diminishes, the human condition.

I owe profound intellectual debt to the diverse communities whose scholarship and practice have laid the foundations upon which these arguments are built. From political theorists who interrogate the meaning of equity, to computer scientists who wrestle with the opacity of algorithms, to civic leaders who strive to bridge digital divides each has contributed threads of insight that have been woven into the conceptual fabric of this book. Their contributions, often glimpsed in passing citations or quiet debates at the margins of academic discourse, have sharpened my understanding and broadened my sense of responsibility.

My deepest gratitude is extended to the editorial team at Springer. Their meticulous care, unyielding commitment to academic integrity, and steady guidance have transformed this manuscript from an exploratory inquiry into a refined scholarly contribution. In a publishing landscape often consumed by speed, their patience and precision remind us that true scholarship requires deliberation and trust.

I am equally indebted to the communities of practice working tirelessly in areas often overlooked data governance, public service accountability, ethical oversight, and inclusive digital infrastructure. Their labors, seldom celebrated, are the very scaffolding upon which equitable societies are built. It is their commitment to the public good that animates the central aspiration of this book: that AI must serve justice before efficiency, transparency before expediency, and humanity before profit.

Finally, to my family and closest confidants, whose silent companionship and enduring encouragement sustained me through the solitude of research and writing: though their names do not appear in these chapters, their presence resonates in every page. Their quiet strength, patience, and faith in the worth of this endeavor are the unseen foundations of its completion.

Competing Interests The author declares, with full academic integrity, that no competing financial, professional, or institutional interests have influenced the conception, development, or final articulation of this work. *Empathy in Algorithms: Shaping Ethical AI for a Just and Inclusive Society* has been undertaken as a wholly independent intellectual endeavor, free from corporate sponsorship, governmental directive, or proprietary obligation.

At no stage whether in the ethical framing of artificial intelligence, the construction of analytical models, or the evaluation of societal impacts was external funding sought or received. The arguments advanced, the frameworks elaborated, and the case analyses presented arise solely from the author's autonomous scholarship and civic concern for the intersection of technology, justice, and public trust.

This volume is not the product of commissioned research, nor does it serve the promotional interests of any vendor, policy body, or commercial entity. Every algorithmic framework, ethical reflection, and socio-technical illustration has been chosen for its scholarly significance, not for alignment with external agendas.

Such a declaration is not a perfunctory formality but an affirmation of principle: that knowledge, to retain its authority, must remain unencumbered by hidden interests. Readers may thus approach this book with confidence that its commitments are intellectual rather than transactional, and that its purpose is to advance the ethical and human-centered design of AI in service to society at large.

Empathy in Algorithms: Shaping Ethical AI for a Just and Inclusive Society

This book initiates the discourse of the book series "AI for Human-Centric Cities: Advancing Health, Inclusion, and Urban Well-Being" with a commanding examination of artificial intelligence not as a neutral technological system, but as a profoundly sociotechnical force capable of entrenching or dismantling systemic inequalities in urban governance. It introduces the notion of "algorithmic empathy" a novel conceptual paradigm that reframes AI systems as civic infrastructures, requiring embedded ethical intelligence, fairness constraints, and social accountability.

The book unfolds through a meticulous traversal of the policy life cycle agenda-setting, formulation, implementation, and evaluation detailing how advanced AI architectures can support equitable public policy, while simultaneously introducing risks of bias, opacity, and exclusion. It introduces a series of mathematically grounded AI models that redefine how justice, transparency, and inclusion are computationally encoded into AI decision systems.

In the domain of public policy agenda-setting, the book presents a Transformer-based Joint Attention Model, wherein multi-head attention is sentiment-weighted to prioritize socially resonant policy themes. It demonstrates how natural language processing (NLP) fused with public sentiment analysis can drive evidence-based governance, ensuring that emergent civic concerns especially those voiced by marginalized groups are algorithmically elevated into decision-making pipelines.

For real-time implementation, the book advances a Contextual Bandit Learning framework, wherein adaptive automation systems such as drones, AI inspectors, or urban service agents optimize interventions under uncertainty through reward-maximizing feedback loops. These models allow AI systems to learn from partial feedback, iteratively refine policy execution, and adapt dynamically to shifting urban conditions.

The evaluation phase of the policy cycle is restructured through a Bayesian Causal Inference model, enabling AI-driven dashboards to infer treatment effects, quantify uncertainty, and perform real-time policy course corrections. The system integrates new data on-the-fly through online Bayesian updates transforming static monitoring into adaptive policy learning.

Critically, the book places ethical AI governance at the structural core of urban AI deployment. Through a combination of group fairness constraints, SHAP-based explainability layers, and audit-ready optimization loss functions, the book introduces an algorithmic architecture of procedural justice where model outcomes are not only accurate but also demographically balanced, interpretable, and open to civic scrutiny.

In public services, the book formalizes a multi-objective deep reinforcement learning (MO-DRL) framework that balances service efficiency and social equity. It equips public administrators with models that optimize service allocation while embedding disparity minimization constraints, ensuring that AI-enhanced public goods do not amplify structural disadvantage.

To address risks of digital exclusion, the book presents Transformer + RLHF-based conversational AI agents, optimized through human feedback signals for contextual accuracy, accessibility, and responsiveness in government-citizen interfaces. A composite reward structure balances linguistic clarity, information correctness, and low system latency ensuring citizen-centered automation.

Labor market transformations are analyzed through a suite of macro-level reinforcement learning and elasticity-based models including a Dynamic Sectoral Reinforcement Learning Model (DSRLM) and a Skills Adjustment Friction Model (SAFM). These models simulate technological displacement, sectoral adaptation, and labor mobility under AI-induced change. Importantly, they quantify skill mismatches and guide targeted reskilling policies through DRL-PPO optimization strategies.

Finally, a scenario-based foresight model powered by Mixture Density Networks (MDNs) supports dynamic reskilling investments, forecasting future skill demand across multiple AI trajectories. This probabilistic model enables governments to anticipate divergent labor futures and prepare human capital strategies that are both agile and inclusive.

About This Book

The book *Empathy in Algorithms: Shaping Ethical AI for a Just and Inclusive Society* part of the book series "AI for Human-Centric Cities: Advancing Health, Inclusion, and Urban Well-Being" explores the foundational role of ethical artificial intelligence in shaping equitable, inclusive, and socially responsive urban governance. Introducing the paradigm of "algorithmic empathy," the book redefines AI not as a neutral technological instrument but as a normative infrastructure capable of reinforcing or redressing structural inequalities. It systematically maps the integration of fairness-aware AI architectures into every stage of the public policy cycle—from sentiment-weighted agenda-setting using transformer-based attention models, to adaptive service deployment via Contextual Bandits, to real-time policy evaluation through Bayesian Causal Inference. The book advances a multi-layered computational framework where deep reinforcement learning, SHAP explainability, and disparity-aware optimization jointly produce intelligent systems that are procedurally just, interpretable, and democratically accountable. Labor market dynamics are addressed through macro-level DRL simulations, forecasting AI-driven job displacement and guiding targeted reskilling investments via mixture density networks. This book offers not only a technical blueprint but also an ethical foundation for governing AI in the public domain, where inclusion, transparency, and human dignity are embedded into the very logic of machine reasoning.

Contents

About the Author

Dr. Dimitrios Sargiotis is a Faculty Dr. and Deputy Director of the PC-Laboratories within the Digital Technology Laboratory for Technical Projects at the School of Civil Engineering, National Technical University of Athens. With more than three decades of academic and professional experience, he stands at the intersection of Artificial Intelligence, Computer Science, Data Governance, and Civil Engineering, pioneering the digital transformation of infrastructures, cities, and societies.

He has authored seventeen scholarly books, one hundred thirty-seven book chapters, and over fifty scientific publications, articles, research reports, and other scholarly contributions disseminated through Springer and other prestigious international outlets, establishing himself as one of the leading voices in AI for Smart Cities, Digital Twins, Intelligent Transportation Systems, Green and Sustainable Infrastructure, and Human-Centric AI. His frameworks emphasize ethics, explainability, and governance, aligning technological innovation with public value and societal well-being.

Dr. Sargiotis provides leadership in European Commission–funded research projects, serves as a reviewer and program committee member for international journals and conferences, and collaborates with esteemed organizations, including Springer, the Programming Research Group of Paris, and the Union of Computer Engineers and Communications of Greece.

Beyond academia, he is an advocate of life-long learning, physical well-being, and cultural engagement, embodying a vision where scientific rigor, ethical responsibility, and human creativity converge. His career demonstrates how Artificial Intelligence can act as a catalyst for sustainable development, social justice, and the flourishing of digitally governed societies worldwide.

Abbreviations

AI Artificial Intelligence machine-based systems capable of learning, reasoning, and decision-making, here positioned as a civic and ethical infrastructure for public policy, governance, and social inclusion.

CNN Convolutional Neural Network deep learning model designed for spatial and visual recognition, used in monitoring, surveillance, and interpretability within civic AI systems.

DBN Deep Belief Network a generative model of layered stochastic variables used for learning latent patterns in social and infrastructural datasets.

DQN Deep Q-Network a DRL method combining Q-learning with deep neural networks to approximate value functions for decision-making under uncertainty.

DRL Deep Reinforcement Learning integration of reinforcement learning with deep neural networks, enabling sequential decision-making in complex, high-dimensional urban governance and service contexts.

EV Electric Vehicle transportation systems powered partly or wholly by electricity, referenced as sustainable infrastructure in AI-driven urban planning.

FFT Fast Fourier Transform algorithm for efficient frequency-domain analysis, applied to cyclical policy data, governance signals, and service demand.

HSR High-Speed Rail modern intercity rail systems embedded within AI-augmented mobility planning as sustainable, multimodal infrastructure.

ITS Intelligent Transportation Systems integrated applications of AI, sensing, and communication to optimize efficiency, safety, and inclusivity in urban mobility.

LSTM Long Short-Term Memory a recurrent neural network (RNN) variant capable of learning long-term dependencies, applied in sequential policy analysis and forecasting service demand.

LWR Lighthill–Whitham–Richards model classical traffic flow theory extended within the book's AI frameworks for systemic policy optimization.

MARL Multi-Agent Reinforcement Learning framework where multiple AI agents cooperate or compete within public systems, enabling fairness-aware governance.
MPC Model Predictive Control advanced optimization technique that forecasts and adjusts system trajectories over a prediction horizon, applied in governance and policy regulation contexts.
NFD Network Fundamental Diagram graphical representation of network flow versus density, reframed as a governance tool in AI-based service allocation.
NLP Natural Language Processing AI methodology for analyzing human language, central to agenda-setting and policy discourse analysis in the book.
PPO Proximal Policy Optimization a stable policy-gradient algorithm used within DRL for training agents in uncertain and dynamic environments such as policy simulations and public service delivery.
RL Reinforcement Learning a paradigm in which agents learn optimal actions through trial-and-error feedback, applied in adaptive policy execution, public service optimization, and governance frameworks.
RNN Recurrent Neural Network neural architecture designed for sequence modeling, employed for temporal dynamics in service delivery and equity-based forecasting.
SHAP SHapley Additive exPlanations interpretable AI method attributing feature contributions to model outputs, ensuring fairness and transparency in AI-driven governance.
TFO Traffic Flow Optimization domain of AI-assisted decision science used metaphorically and technically to enhance governance efficiency and equity.

Annotations

Artificial Intelligence (AI)	Computational systems capable of learning, reasoning, and decision-making; in this book framed as civic infrastructure with ethical, social, and governance implications.
Public Policy	The processes of agenda-setting, formulation, implementation, and evaluation where AI is embedded to enhance transparency, equity, and efficiency.
Ethical AI	Denotes AI systems explicitly designed with fairness, accountability, transparency, and inclusivity at their core.
Fairness	A recurring principle ensuring AI does not exacerbate systemic inequalities, operationalized through fairness-constrained optimization and group equity constraints.
Transparency	The requirement that AI decisions and processes remain understandable and open to scrutiny, supporting accountability and public trust.
Accountability	The obligation of AI systems and their developers to provide traceable, auditable, and justifiable decision-making structures.
Governance	Frameworks and regulatory mechanisms to oversee AI deployment, ensuring compliance with democratic, ethical, and legal standards.
Inclusion/Inclusivity	The commitment to design AI that benefits marginalized populations and prevents digital exclusion, reinforcing social equity.
Digital Divide	Refers to the gap in access to technology and AI benefits across socioeconomic and geographic groups, a central concern in the book.

Bias	Systematic distortions in AI outputs due to data, algorithms, or design, addressed through bias detection, audits, and mitigation strategies.
Privacy	The protection of personal and sensitive data in AI-driven public services, emphasized against risks such as surveillance and inference attacks.
Equity	Principle of ensuring fair distribution of AI benefits, distinct from equality, embedded into "equity-by-design" frameworks.
Social Impact	The societal consequences of AI integration in governance and services, requiring evaluation beyond technical efficiency.
Reskilling	Strategies for workforce transition in response to technological unemployment risks induced by automation and AI adoption.
Human-Centered AI (HCAI)	An approach that prioritizes human needs, rights, and values in the design and deployment of AI systems.
Procedural Justice	A framework where AI outcomes are not only accurate but also demographically fair, interpretable, and open to civic review.
Surveillance	The monitoring capabilities of AI in public services, raising ethical debates around privacy, consent, and accountability.
Algorithmic Empathy	A novel concept introduced in the book: reframing AI as capable of empathetic response to human and civic concerns through ethical embedding.
Trust	Essential social capital for AI adoption, linked to explainability, transparency, and compliance with ethical governance.
Regulatory Frameworks	Legal and institutional guidelines that govern the ethical and responsible deployment of AI in urban and public systems.

Chapter 1
Artificial Intelligence and Public Policy

This chapter examines how artificial intelligence (AI) is fundamentally reshaping the architecture of public policy, from the earliest moments of agenda-setting to the long horizon of policy evaluation. It charts a journey through the entire policy cycle, showing how AI's analytical, predictive, and adaptive capabilities infuse each stage with unprecedented speed, depth, and precision. At the outset, advanced natural language models sift through torrents of social data, translating the collective pulse of society into actionable intelligence. As policy choices move toward implementation, adaptive learning systems and automation frameworks transform governance into a real-time, responsive enterprise, able to adjust to shifting conditions and dynamic contexts. Evaluation, once a retrospective exercise, becomes a forward-looking science powered by causal inference and continuous feedback loops.

Yet this chapter does not restrict itself to technical innovation alone. It confronts the ethical frontiers of AI in governance, accountability, transparency, and fairness, introducing rigorous mathematical frameworks to ensure that efficiency does not come at the expense of equity. Case studies from multiple continents illustrate how governments, agencies, and institutions are already harnessing AI to improve service delivery, enhance citizen trust, and address systemic inequalities. At the same time, the text tackles one of the most pressing challenges of our time: the impact of AI on labor markets. By examining the dual forces of displacement and task creation, it frames strategies for workforce transition, reskilling, and inclusive adaptation.

This chapter presents AI not simply as a tool of administration but as a transformative force in democratic governance, a catalyst for reimagining the social contract in the digital century.

D. Sargiotis, *Empathy in Algorithms*, https://doi.org/10.1007/978-3-032-32128-2_1

1.1 Introduction

Artificial intelligence can create, intensify, or institutionalize ethical and justice-related problems when algorithmic systems transform data into evidence, evidence into decisions, and decisions into consequential forms of social action. Tsamados et al. (2022) show that algorithms are now embedded in major social infrastructures, including education, healthcare, finance, courts, local government, and national governance, and therefore increasingly participate in decisions that affect individuals and populations. Their analysis stresses that algorithms are not ethically neutral, since algorithmic systems may encode gendered associations in search and translation, reproduce discriminatory patterns in advertising, or amplify unequal treatment in healthcare when predictive models are trained on data shaped by prior social inequalities (Tsamados et al. 2022).

To clarify these risks, Tsamados et al. (2022) identify six ethical concerns: inconclusive evidence, inscrutable evidence, misguided evidence, unfair outcomes, transformative effects, and traceability. These concerns indicate that the ethical problem of AI is not reducible to technical bias alone; it also involves uncertainty in algorithmic inference, opacity in model operation, flawed or incomplete evidence, discriminatory outcomes, wider social consequences, and difficulties in attributing responsibility across the design, development, deployment, and institutional use of algorithmic systems (Tsamados et al. 2022). Rafanelli (2022) extends this discussion by arguing that AI raises questions of justice because it is not a replacement for human power, but a new way of exercising it.

Drawing on political theory and philosophy, she explains that justice concerns the fair adjudication of claims to liberties, opportunities, resources, and modes of treatment, while AI systems may either promote justice or entrench injustice depending on how they are designed, trained, and used (Rafanelli 2022). Her examples of Amazon's abandoned résumé screener, Google's photo sorter, and facial-recognition systems with uneven performance across gender and skin tone show that AI can facilitate institutional discrimination, structural injustice, and epistemic injustice by automating historical prejudice, reinforcing unequal social structures, and shaping whose experiences are recognized as credible or intelligible (Rafanelli 2022).

Wegner et al. (2026) further refine this conceptual foundation by arguing that algorithmic fairness research has often been dominated by distributive approaches, which concentrate on outcomes, statistical metrics, and bias mitigation, but may insufficiently address power relations, domination, oppression, marginalization, and mutual respect. Their systematic review identifies an emerging relational and structural turn in algorithmic fairness research, according to which fairness cannot be understood only as the equal distribution of algorithmic outputs, but must also be examined through the social relations, institutions, and power asymmetries within which algorithmic systems operate (Wegner et al. 2026).

These studies suggest that the ethical and justice problems created by AI must be defined before technical solutions are proposed: AI systems become ethically consequential when they convert historically conditioned data into apparently objective evidence, when opaque models guide decisions affecting rights or opportunities, when institutional actors treat algorithmic outputs as neutral, and when affected communities lack meaningful power to understand, contest, or reshape the systems that classify and govern them (Tsamados et al. 2022; Rafanelli 2022; Wegner et al. 2026).

How Do Fairness, Bias, Discrimination, and Structural Injustice Shape the Ethical Problem of AI?

Fairness, bias, discrimination, and structural injustice form a connected problem space in contemporary AI ethics because biased systems may translate technical, statistical, or representational distortions into unequal social outcomes. Ferrara (2024) defines AI bias as systematic error in decision-making that can lead to unfair outcomes and shows that such bias may arise from data collection, algorithm design, and human interpretation; machine-learning models may then learn and reproduce these patterns, producing discriminatory effects in domains such as hiring, lending, healthcare, criminal justice, and generative AI media (Ferrara 2024).

The same study emphasizes that biased AI systems can perpetuate systemic discrimination and inequality, reinforce stereotypes, restrict access to essential services, and generate new forms of discrimination based on race, gender, skin color, ethnicity, socioeconomic status, or physical appearance (Ferrara 2024). Jarrahi et al. (2025) complicate this picture by showing that "bias" is not used uniformly across AI research: some communities treat it as a statistical or computational parameter, whereas others understand it as a sociotechnical and ethical problem requiring contextual interpretation. AlMakinah et al. (2025) further support a lifecycle-oriented view of AI bias, presenting bias as a phenomenon that can be mapped, analyzed, and mitigated across AI systems rather than corrected only at a single technical point.

Kasirzadeh (2022) moves the discussion from bias mitigation to structural injustice, arguing that algorithmic fairness research has primarily relied on mathematical metrics rooted in distributive justice, while such metrics may fail to accommodate injustices produced by durable social structures, power relations, actors, and institutional arrangements. On this view, a locally fair algorithmic outcome may still be morally inadequate if it leaves untouched the broader conditions that create and sustain domination, deprivation, or disadvantage (Kasirzadeh 2022). Wegner et al. (2026) extend this critique by showing that algorithmic fairness research has often centered on statistical metrics and distributive outcomes, whereas an emerging relational and structural turn foregrounds domination, oppression, marginalization, mutual respect, and the quality of social relations.

These studies suggest that fairness in AI cannot be reduced to equalizing outputs or removing sensitive variables; rather, it must also examine how AI systems interact with historical inequality, institutional responsibility, sociotechnical context, and structural power.

How Can AI-Supported Public Decisions Remain Transparent, Accountable, and Legitimate?

Transparency, accountability, and legitimacy are interdependent but analytically distinct requirements for the responsible use of AI in public decision-making. De Fine Licht and de Fine Licht (2020) argue that transparency is often treated as a remedy for the "black box" problem of AI, yet they caution that full transparency does not necessarily produce public legitimacy. Their analysis distinguishes between transparency of the final decision, transparency of the decision-making process, and transparency of the rationale behind the decision. In the context of public AI systems, they suggest that legitimacy is more likely to be supported by a limited, justification-focused form of transparency than by unrestricted disclosure of code, data, or model architecture, since the latter may be technically unintelligible, excessive, or even counterproductive for public acceptance.

What matters, therefore, is that affected citizens receive intelligible reasons for AI-supported decisions, including an explanation of the grounds on which the decision was made and the institutional context in which responsibility is assigned (de Fine Licht and de Fine Licht 2020).

Novelli et al. (2024) extend this discussion by defining accountability not as a vague ethical aspiration, but as a structured relation of answerability. In their account, accountability requires authority recognition, interrogation, and limitation of power; it also involves seven features, namely context, range, agent, forum, standards, process, and implications. Applied to AI, this means that responsible governance must specify what the accountability relation concerns, who is accountable, to whom an account is owed, according to which standards conduct is assessed, through which procedures, and with what consequences (Novelli et al. 2024).

Their account is especially relevant because AI systems are sociotechnical systems: harmful or contested outcomes may arise not only from a model but also from training data, design choices, deployment contexts, institutional use, regulatory gaps, or the interaction among these elements. Accountability in AI must therefore include compliance, reporting, oversight, and enforcement mechanisms across the lifecycle of design, development, deployment, and review (Novelli et al. 2024). Langer (2024) brings this problem into the constitutional and administrative domain by arguing that automated decision-making in public administration raises questions of decision-making power, responsibility, and democratic legitimacy. Her analysis of the non-delegation principle shows that public authorities cannot simply transfer rights-affecting decisions to algorithms or private software developers in ways that weaken governmental accountability, reviewability, and contestability.

Because technical choices in automated administration, such as data selection, model design, threshold setting, interface design, and the treatment of uncertain cases, may embed normative judgments, democratic legitimacy requires that public institutions retain meaningful control over delegated decision-making and preserve institutional mechanisms through which affected citizens can question, review, or contest AI-supported decisions (Langer 2024). Taken together, these studies suggest that legitimate AI governance requires more than explainable models or formal transparency. It requires accessible justification, identifiable responsible actors, enforceable accountability structures, institutional oversight, and contestable decision-making pathways through which public power remains democratically controlled even when exercised through algorithmic systems.

How Can Responsible AI Governance Translate Ethical Principles into Accountable Institutional Practice?

Responsible AI governance should be understood as the practical capacity to translate AI principles into organizational structures, lifecycle controls, stakeholder responsibilities, and mechanisms of review. Papagiannidis et al. (2025) argue that, although responsible AI principles have been widely discussed in policies, regulations, and organizational frameworks, there remains limited understanding of how such principles are operationalized in the design, execution, monitoring, and evaluation of AI applications. Their review therefore frames responsible AI governance through structural, relational, and procedural practices, emphasizing that governance is not only a matter of declaring ethical commitments but of embedding them into organizational routines, decision rights, implementation processes, and evaluative mechanisms (Papagiannidis et al. 2025). Batool et al. (2024) similarly present responsible AI governance as a response to the risks generated by the increasing deployment of AI across domains, especially risks related to ethics, transparency, bias, and fairness. Their systematic literature review organizes the field around four central questions: who is accountable for AI systems, what elements of AI are governed, where governance occurs within the AI development lifecycle, and how governance is executed through mechanisms such as frameworks, tools, standards, policies, and models (Batool et al. 2024). In this sense, responsible AI governance requires a lifecycle-oriented architecture that identifies accountable actors, specifies the objects and stages of governance, and connects ethical principles to concrete procedures for design, deployment, monitoring, and correction.

Ethics-based AI auditing provides the assurance layer through which responsible AI governance becomes observable, assessable, and institutionally actionable. Laine et al. (2024) show that AI auditing has emerged as a response to risks such as algorithmic opacity, unethical conduct, unintended consequences, and discriminatory bias. Their systematic review explains that ethics-based AI auditing engages with principles such as fairness, transparency, non-maleficence, responsibility, privacy, trust, beneficence, and autonomy, while also noting that these principles are conceptualized differently across social, technical, process-oriented, and outcome-oriented dimensions (Laine et al. 2024). The same study identifies system developers and deployers, the wider public, researchers, auditors, AI users, and regulators as major stakeholders, and argues that AI auditing contributes through guidance, methods, tools, frameworks, awareness, and empowerment (Laine et al. 2024). This makes auditing more than a compliance ritual: it becomes a structured process for gathering evidence, examining risks and controls, evaluating conformity with standards and requirements, and supporting accountability across the AI system lifecycle. Laine et al. (2024) also distinguish between internal and external AI auditing, noting that internal auditing can access system information not normally available to outsiders, while external auditing often relies on public information and system outputs; both forms may therefore be necessary for a comprehensive assessment of AI systems.

Data governance forms the foundational layer beneath both responsible AI governance and AI auditing, because the quality, integrity, security, privacy, and stewardship of data shape the reliability and accountability of AI systems before any model is trained or deployed. Sargiotis (2024) defines data governance as the overall management of the availability, usability, integrity, and security of organizational data, and presents it as a framework concerned with data quality, data security, data privacy, data management, policies, standards, roles, responsibilities, stewardship, ownership, compliance, auditing, maturity assessment, and continuous improvement (Sargiotis 2024). This is directly relevant to responsible AI, since AI systems cannot be governed responsibly if their data foundations are inaccurate, incomplete, insecure, poorly stewarded, or non-compliant. Sargiotis (2024) further emphasizes that data governance must include monitoring, auditing, accountability mechanisms, policy enforcement, stakeholder involvement, training, and continuous improvement. Consequently, AI governance may remain incomplete if models are audited only at the output level while upstream conditions such as data quality, provenance, access control, privacy compliance, security, and stewardship remain insufficiently governed.

Responsible AI requires a layered governance architecture: data governance secures the informational foundation, responsible AI governance defines the institutional and lifecycle mechanisms, and ethics-based auditing evaluates whether AI systems actually conform to ethical, technical, legal, and organizational expectations.

How Do Data-Driven Systems Transform Surveillance and Exclusion into Problems of Data Justice?

Data justice, surveillance, and exclusion are interconnected because data-driven systems do not merely record social life; they make individuals and groups visible, measurable, sortable, and governable in institutionally consequential ways. Dencik and Sanchez-Monedero (2022) define data justice as a framework for examining the relationship between datafication and society through an explicit concern with social justice, thereby moving the debate beyond digital rights alone toward the broader political question of what is at stake when social relations are translated into data. Taylor et al. (2025) extend this argument into the field of AI governance by arguing that governing artificial intelligence also requires governing data, since AI systems depend on data infrastructures, statistical modeling, large-scale data flows, and the political and commercial interests embedded in those systems. Their analysis suggests that AI intensifies the risks of datafication by making categorization and sorting more dynamic, probabilistic, and infrastructural, while also creating new concentrations of commercial and political power. For this reason, they propose that just governance at the intersection of data and AI should be assessed through four benchmarks: the preservation and strengthening of public infrastructures and public goods, inclusiveness, contestability and accountability, and global responsibility.

Surveillance is a central data justice concern because the collection, integration, and analysis of data can shape the opportunities, constraints, and forms of scrutiny to which different populations are exposed. Dencik et al. (2016) argue that the Snowden leaks revealed extensive state-corporate surveillance and showed how everyday communication had become embedded in regimes of control based on the datafication of social life. They further argue that anti-surveillance resistance should not remain confined to techno-legal responses such as encryption, anonymity, privacy, and data protection, because data-driven surveillance also raises broader social justice questions concerning citizenship, economic inequality, welfare, discrimination, and the political organization of society. Brayne (2017), in an empirical study of the Los Angeles Police Department, shows how big data analytics transform surveillance practices by supplementing discretionary risk assessments with quantified risk scores, shifting data use toward predictive rather than reactive or explanatory purposes, enabling automated alerts across large populations, lowering thresholds for inclusion in law-enforcement databases, and merging previously separate datasets, including data from non-criminal-justice institutions. These transformations matter for justice because they allow surveillance to become wider, deeper, more predictive, and more institutionally portable, with consequences for law, inequality, and organizational practice.

Exclusion emerges when data systems classify people in ways that restrict access, intensify stigma, or reshape how public institutions identify risk, vulnerability, suspicion, or eligibility. Dencik et al. (2018) show that citizen scoring in UK public services combines data from multiple sources to categorize citizens, allocate services, and predict behavior. Their report documents applications in fraud detection, public health, child welfare, social services, and policing, while also identifying concerns about extensive data collection and sharing, sensitive council-held data, biased decision-making based on historically skewed datasets, targeting and stereotyping through risk labels, limited transparency, weak public knowledge, inadequate consent, and insufficient oversight. These concerns directly connect data analytics to exclusion because citizens may be sorted into institutional categories that affect access to support, exposure to intervention, or suspicion of wrongdoing without a clear understanding of how such classifications are produced, governed, or contested (Dencik et al. 2018).

At this point, data security and privacy become relevant as governance safeguards, although they do not replace the broader theory of data justice. Sargiotis (2024) frames data security and privacy as critical components of data governance, emphasizing the protection of sensitive information through measures such as encryption, access control, data masking, privacy by design, regulatory compliance, incident response planning, monitoring, audits, and the balancing of data accessibility with security and privacy. This perspective is relevant to data justice, surveillance, and exclusion because systems that classify, monitor, or profile people require not only conceptual scrutiny but also enforceable governance controls over how data is collected, accessed, protected, shared, retained, audited, and used. In this sense, Sargiotis (2024) supplies a practical governance bridge: data justice identifies the social and political harms of datafication, while data security and privacy practices help establish the institutional conditions under which data-driven systems may become more accountable, less intrusive, and less vulnerable to misuse (Sargiotis 2024).

Data justice requires more than privacy protection or technical data management. It requires scrutiny of how data-driven and AI-enabled systems classify populations, expand surveillance, redistribute institutional attention, and produce forms of inclusion or exclusion. A justice-oriented approach must therefore ask who is made visible, who is rendered risky or suspicious, who benefits from data-driven governance, who bears its burdens, and what mechanisms exist for accountability, contestation, refusal, secure data handling, and democratic participation.

Table. 1.1 establishes the normative ground on which the following technical models should be read. The mathematical frameworks introduced in the subsequent sections are not presented as definitive solutions to ethical and justice problems, but as operational translations of selected governance commitments. Fairness constraints, explainability functions, auditing mechanisms, optimization objectives, and privacy-preserving safeguards can support responsible AI only when they remain grounded in prior conceptual clarification, institutional accountability, and democratic contestability. In this sense, the models that follow demonstrate how AI can respond to these questions in operational terms, while leaving space for ethical reasoning, public oversight, and political judgment.

Artificial intelligence (AI) is playing an increasingly significant role in shaping public policy, fundamentally transforming the way decisions are made and implemented. The integration of AI into public policy-making processes is seen across various phases of the policy cycle, including agenda-setting, policy formulation, decision-making, implementation, and evaluation.

In agenda-setting, AI-driven tools such as natural language processing (NLP) are being used to analyze large datasets, including social media trends, public sentiment, and media content, to help policymakers identify emerging issues and prioritize public concerns. AI techniques allow policymakers to frame policy problems more efficiently by extracting critical insights from vast amounts of data. This process, often referred to as evidence-based policymaking, emphasizes the importance of using real-time data to address societal challenges more effectively (Yar et al. 2024).

To support real-time agenda-setting in public policy, AI models must be capable of interpreting complex, high-volume public discourse across multiple domains. A powerful approach to this task is the transformer-based joint attention model, which combines semantic topic modeling with sentiment-weighted attention to detect emergent issues and rank them by social relevance. At the core of this model is the multi-head self-attention function (semantic-sentiment attention mapping for policy issue extraction):

$$Attention(Q, K, V) = softmax\left(\frac{QK^{\mathrm{T}}}{\sqrt{d_k}} + S\right)V$$

Table 1.1 Conceptual roadmap for AI ethics and justice

Guiding question	Core ethical/justice problem	Main conceptual emphasis	Governance implication
How do fairness, bias, discrimination, and structural injustice shape the ethical problem of AI?	Bias, discrimination, and structural injustice	Fairness is not reducible to statistical parity, equalized outputs, or bias correction alone; it must also account for sociotechnical context and structural power	AI systems require sociotechnical fairness assessment, institutional accountability, and continuous review of discriminatory effects
How can AI-supported public decisions remain transparent, accountable, and legitimate?	Opacity, accountability gaps, and legitimacy deficits in public decision-making	Transparency must provide intelligible justification and contextual explanation, not merely technical disclosure of code, data, or model architecture	Public AI requires identifiable responsibility, reviewability, contestability, oversight, and mechanisms through which affected citizens can question decisions
How can responsible AI governance translate ethical principles into accountable institutional practice?	Gap between ethical principles and operational governance	Responsible AI depends on organizational structures, procedural controls, stakeholder responsibilities, lifecycle monitoring, and auditing mechanisms	Governance must include standards, decision rights, lifecycle controls, ethics-based auditing, compliance mechanisms, and enforceable review procedures
How do data-driven systems transform surveillance and exclusion into problems of data justice?	Surveillance, profiling, exclusion, data-driven sorting, and unequal institutional visibility	Data systems make people visible, measurable, sortable, and governable; data justice therefore requires attention to classification, participation, and power	Data justice requires accountable data governance, secure data handling, public participation, contestability, refusal where appropriate, and democratic oversight

where: Q, K, V are the query, key, and value matrices derived from input token embeddings (from social media, news, or legislative transcripts), d_k is the key dimension, and S is a sentiment-modulated score matrix derived from contextual polarity analysis. This formulation enhances classical transformer attention by amplifying or attenuating token weights based on public sentiment, allowing the model to prioritize emotionally resonant or contentious policy topics.

Following attention, the model uses a joint embedding function (topic-sentiment fusion layer), which unifies semantic and affective vectors to generate a robust representation of emerging policy discourse:

$$z_i = \gamma \cdot t_i + (1 - \gamma) \cdot s_i$$

where: z_i is the fused representation of document i, t_i is the topic embedding vector (learned via contextualized token clustering), s_i is the sentiment vector (learned via affective lexicons or fine-tuned classification), and $\gamma \in [0, 1]$ is a trainable parameter balancing semantic and emotional content. This fusion ensures that topics with both high relevance and high emotional salience are elevated in the policy prioritization output.

To generate actionable outputs, the model applies a policy agenda ranking function (public issue prioritization layer):

$$r_i = w^T z_i + b$$

where: r_i is the predicted policy priority score for topic i, w is a learned weight vector, and b is a bias term. This final scalar guides agenda-setting by allowing policymakers to rank and categorize emerging topics by urgency, societal relevance, or political feasibility.

This transformer-based joint attention model transforms raw public discourse into structured policy intelligence. By simultaneously processing topical relevance and sentiment weight, it allows AI systems to assist policymakers in detecting latent issues, framing them based on public resonance, and prioritizing them for real-time agenda formulation. The architecture is inherently adaptable across domains from healthcare and climate policy to urban safety and digital governance and supports the goals of evidence-based policymaking. It ensures that public sentiment and social narratives are not only heard but computationally integrated into the formation of equitable, responsive, and data-driven policy agendas.

Framework: Transformer-Based Policy Agenda Setting Framework (TPASF)
Layered AI Architecture for Real-Time Evidence-Based Policymaking

Layer	Main functionality	AI module(s) and components	Operational scope
Layer 1	Public discourse collection and tokenization	Data ingestion and preprocessing module	Raw media content, social media posts → Text preprocessing and tokenization → Structured input embeddings
Layer 2	Semantic-sentiment attention mapping	Transformer with sentiment-augmented attention	Token embeddings → Sentiment-modulated multi-head attention → Contextualized representations
Layer 3	Semantic-emotional fusion embedding	Topic-sentiment fusion layer	Contextual and affective embeddings → Vector interpolation → Unified discourse representation

(continued)

(continued)

Layer	Main functionality	AI module(s) and components	Operational scope
Layer 4	Public issue prioritization	Policy agenda ranking layer	Fused representations → Linear transformation → Ranked topic priorities
Layer 5	Human-in-the-loop policy review	Collaborative policy validation and governance module	Ranked issue list → Policy analyst feedback → Decision refinement and validation

AI's influence extends into the implementation phase as well, where automation tools such as robotics, drones, and ML techniques streamline the process. These technologies have been used in law enforcement, healthcare, and tax administration, improving efficiency and reducing costs. AI can also monitor real-time data to provide feedback during policy execution, suggesting necessary interventions or adjustments to enhance effectiveness (Visvizi 2022). During the implementation phase of urban projects, AI's operational influence becomes both immediate and dynamic. Systems must respond in real time to continuously changing conditions adjusting strategies based on sensor inputs, resource availability, or human feedback. A computationally efficient method for this challenge is the Contextual Bandit Learning framework, which supports real-time adaptive automation through selective decision-making and feedback loops. Unlike traditional supervised models, this AI structure operates with partial feedback, ideal for scenarios such as robotic construction, drone monitoring, or adaptive enforcement policies where only one action's outcome is observed per iteration.

At the core of this approach is the contextual reward maximization function (real-time decision optimization in bandit systems):

$$a_t = \arg\max_{a \in \mathcal{A}} \mu_a(\mathrm{x}_t)$$

where: a_t is the action chosen at time t (e.g., deploy a drone to Sector$A$, trigger robotic inspection at Site B), $\mathcal{A}$ is the set of available actions, x_t is the observed context at time t (such as weather, sensor input, or policy deviation), and $\mu_a(\mathrm{x}_t)$ is the expected reward (or effectiveness) of action aa given the context x_t. This formulation allows the system to learn, at every time step, which action maximizes performance under current implementation conditions, even without full knowledge of all possible outcomes.

To ensure long-term learning and robust policy adaptation, the model uses a regret minimization criterion (efficiency objective over time), comparing the cumulative performance of the AI's decisions to the optimal theoretical policy:

$$R_T = \sum_{t=1}^{T} \left[\mu_{a^*}(\mathrm{x}_t) - \mu_{a_t}(\mathrm{x}_t) \right]$$

where: R_T is the total regret over T rounds, a^* is the optimal action in hindsight for each context x_t, and $\mu_{a_t}(x_t)$ is the expected reward of the chosen action. Minimizing regret ensures the model converges to near-optimal policies while maintaining adaptability under uncertainty.

To make implementation AI systems resilient to anomalies or policy misalignment, we introduce a corrective intervention signal (real-time feedback mechanism), which modifies action priorities based on outcome deviation:

$$\delta_t = \hat{y}_t - y_t, \; x_{t+1} \leftarrow x_t + \eta \cdot \delta_t$$

where: $\hat{y}_t$ is the predicted outcome, y_t is the observed result (e.g., actual energy use, drone success rate), δ_t is the prediction error, and η is the learning rate. This update rule refines future context representations, enabling the AI to adjust to unexpected implementation realities while preserving performance.

This Contextual Bandit framework equips urban automation systems with the ability to monitor, adapt, and refine actions during implementation learning from feedback without requiring full retraining. It supports intelligent drones, autonomous inspection robots, and policy recommendation agents that can evaluate performance in real time and adjust behavior accordingly. Crucially, the model's partial-feedback design reflects the on-the-ground constraints of real cities, where outcomes are often observable only after intervention. In doing so, it translates AI's potential into tangible, continuous implementation efficiency making city management not only smarter, but also more responsive, self-correcting, and cost-effective.

Framework: Contextual Bandit-Based Urban Implementation Framework (CBUIF)
Layered AI Architecture for Real-Time Policy Execution and Feedback

Layer	Main functionality	AI module(s) & components	Operational scope
Layer 1	Input signal processing and environmental encoding	Real-time context encoding and stream processing module	Raw sensor data, policy execution logs → Feature extraction and normalization → Context vectors for AI decision-making
Layer 2	Real-time decision making via action selection	Contextual bandit action selection	Context vectors → Reward estimation and action scoring → Selected action per iteration
Layer 3	Performance evaluation and learning adjustment	Cumulative regret tracking	Chosen action and contextual reward → Regret calculation → Policy improvement
Layer 4	Online model correction with feedback	Online contextual bandit parameter update	Predicted vs actual results → Compute feedback error → Adjust context representation

(continued)

(continued)

Layer	Main functionality	AI module(s) & components	Operational scope
Layer 5	Operator supervision and adaptive reprioritization	Human-in-the-loop orchestration and reprioritization module	Updated context and performance reports → Human-in-the-loop review → Adjusted policy directives

Finally, the evaluation phase of the policy cycle benefits from AI's capacity to analyze policy impacts through real-time data simulations. AI-driven dashboards can monitor policy outcomes, providing timely feedback to decision-makers for future adjustments. For instance, AI has been used in climate policy and public health to assess the effectiveness of interventions, allowing for dynamic policy readjustments as new data becomes available (Yar et al. 2024). In the final stage of the policy cycle evaluation AI plays a transformative role by enabling governments to assess not just what is happening, but why it is happening. A robust approach for this is the Bayesian Causal Inference framework, which estimates the causal effect of policies in real time, enriched with online learning mechanisms to incorporate new evidence and adapt predictions dynamically. This model allows AI-driven dashboards to go beyond correlation-based monitoring, instead inferring how interventions have shaped outcomes under evolving conditions. The foundational component is the posterior estimation function (Bayesian causal treatment effect estimator):

$$P(\tau|\mathcal{D}) = \frac{P(\mathcal{D}|\tau)P(\tau)}{P(\mathcal{D})}$$

where: τ represents the treatment effect (e.g., impact of a climate subsidy or vaccination campaign), $\mathcal{D}$ is the observed data, $P(\mathcal{D}|\tau)$ is the likelihood of the data given the treatment effect, $P(\tau)$ is the prior belief about the treatment's impact, and $P(\mathcal{D})$ is the evidence (marginal probability of the data). This Bayesian formulation allows the model to express not only effect magnitude but also uncertainty critical for high-stakes policy evaluation.

To adapt as new data becomes available, the system employs an online Bayesian update rule (real-time inference refinement):

$$P_{t+1}(\tau) \propto P_t(\tau) \cdot P(\mathcal{D}_{t+1}|\tau)$$

where: $P_t(\tau)$ is the current belief distribution over policy impact at time t, and $\mathcal{D}_{t+1}$ is the newly observed data. This formulation enables continuous learning and updating, ensuring that as new interventions are implemented or external conditions change, the system re-evaluates effectiveness without retraining from scratch.

The output of the model is integrated into a real-time policy feedback function (dashboard decision support model):

$$\hat{y}_{t+1} = f(\tau_t, \mathcal{C}_t, \mathrm{X}_t)$$

where: $\hat{y}_{t+1}$ is the predicted outcome for the next policy window (e.g., carbon reduction, infection rate), τ_t is the estimated causal effect at time t, C_t represents external contextual variables (e.g., economic shifts, behavioral responses), and X_t 0 is the matrix of observed covariates. This function enables policymakers to view projected impacts, identify lagged effects, and preemptively adjust strategies.

This Bayesian Causal Inference model with online updating transforms traditional policy evaluation into a dynamic, real-time learning system. It provides both quantitative impact estimates and uncertainty bounds, supporting more nuanced and transparent decision-making. By embedding causal reasoning directly into AI-driven dashboards, governments gain the ability to continuously assess whether a policy is working, for whom, and under what conditions enabling targeted refinements without delay. This model exemplifies how AI can fulfill the promise of adaptive governance: evidence-rich, temporally responsive, and grounded in computational rigor.

Framework: Bayesian Causal Policy Evaluation Framework (BCPEF)
Layered AI Architecture for Real-Time Impact Assessment and Adaptive Governance

Layer	Main functionality	AI module(s) & components	Operational scope
Layer 1	Data stream integration and covariate structuring	Real-time context encoding and stream processing module	Policy outcome data, behavioral indicators → Preprocessing and alignment → Time-indexed observational matrix
Layer 2	Causal effect estimation via posterior inference	Bayesian posterior inference	Prior distributions and observational data → Posterior treatment effect estimation → Probabilistic causal inference
Layer 3	Dynamic learning via online evidence update	Sequential Bayesian update	Posterior from previous cycle and new data → Distribution refinement → Updated belief over policy effect
Layer 4	Dashboard-based forecasting of policy impact	Causal predictive dashboard function	Causal estimates, covariates, context variables → Outcome prediction → Projected impact trajectory
Layer 5	Decision-maker review and strategic adjustment	Human-in-the-loop orchestration and reprioritization module	Impact projections and uncertainty bounds → Human oversight and calibration → Recalibrated policy directions

However, the integration of AI in public policy development also presents several challenges, particularly in terms of ethics, accountability, and transparency. Issues such as data privacy, algorithmic biases, and the risk of unfair decision-making have been widely acknowledged. There is a growing need for robust governance frameworks that can ensure the ethical use of AI in public policy, addressing concerns related to the potential for AI to exacerbate existing social inequalities (Visvizi 2022). While AI has shown immense potential in enhancing the speed, efficiency, and responsiveness of public policymaking, its integration must be governed by a framework that ensures ethical compliance, algorithmic accountability, and equity in decision outcomes. A mathematically grounded approach to such governance is the use of fairness-constrained policy learning, embedded within an auditable decision architecture that enables transparency at every computational layer. This framework provides a formal mechanism to detect, quantify, and mitigate potential algorithmic harms, particularly those that may reinforce or amplify social inequities.

At the foundation of this approach is a policy fairness constraint (group fairness regularization objective), ensuring that predicted outcomes remain balanced across sensitive demographic subgroups:

$$\mathcal{L}_{\text{fair}} = \left|\mathbb{E}_{x\in\mathcal{G}_0}[\hat{y}] - \mathbb{E}_{x\in\mathcal{G}_1}[\hat{y}]\right|$$

where: $\mathcal{G}_0$ and $\mathcal{G}_1$ are two population subgroups defined by a sensitive attribute (e.g., gender, income, ethnicity), and $\hat{y}$ is the predicted policy outcome (e.g., benefit eligibility, resource allocation). This formulation minimizes disparities in outcomes, thereby promoting procedural fairness.

To embed transparency into algorithmic workflows, we introduce an explainable decision decomposition (traceability mechanism for auditability):

$$\hat{y} = f_\theta(x) = \sum_{k=1}^{K} \phi_k \cdot x_k$$

where: $f_\theta(x)$ is the AI model's prediction, and each ϕ_k is a learned contribution score (e.g., via SHAP- *SHapley Additive exPlanations* or LIME-*Local Interpretable Model-Agnostic Explanations,* explainability models) indicating how input variable x_k influenced the output. This decomposition allows stakeholders policymakers, auditors, and citizens to understand and scrutinize how decisions are formed.

The total model objective becomes a regularized governance loss function (ethical optimization under accountability constraints):

$$\mathcal{L}_{\text{total}} = \mathcal{L}_{\text{task}} + \lambda \cdot \mathcal{L}_{\text{fair}} + \beta \cdot \mathcal{L}_{\text{explain}}$$

where: $\mathcal{L}_{\text{task}}$ is the original predictive or classification loss (e.g., policy impact prediction), $\mathcal{L}_{\text{fair}}$ enforces fairness constraints, and $\mathcal{L}_{\text{explain}}$ quantifies explainability deviation (e.g., model complexity or feature interpretability). The coefficients λ and β calibrate the trade-offs between accuracy, fairness, and interpretability.

This fairness-constrained, audit-ready model provides a formalized pathway to ethically govern AI systems deployed within the policy domain. By embedding fairness constraints directly into the model's objective and enforcing interpretability through decomposable structures, it ensures that AI systems respect both human rights and democratic transparency. In practice, this approach empowers stakeholders to evaluate not only what an AI system predicts, but also why it predicts it and whether it does so fairly. As cities and states increasingly adopt AI for policy design and implementation, such governance frameworks are essential for preventing algorithmic discrimination, safeguarding public trust, and ensuring that technological progress enhances not undermines social justice.

Framework: Ethical Governance-Aware Ai Policy Framework (EGAPF)
Layered AI Architecture for Fairness-Constrained and Explainable Decision-Making

Layer	Main functionality	AI module(s) & components	Operational scope
Layer 1	Sensitive attribute encoding and audit baseline structuring	Fairness-aware data preparation and encoding module	Policy datasets, demographic attributes → Preprocessing and subgroup assignment → Structured fairness-aware data
Layer 2	Fairness-constrained outcome optimization	Group fairness regularizer	Subgroup predictions → Compute inter-group disparity → Penalize outcome imbalance
Layer 3	Model interpretability via attribution decomposition	Additive feature attribution predictor	Input features and trained model → Feature contribution scores → Explanation of prediction logic
Layer 4	Unified ethical optimization objective	Fairness-and-explainability regularized learning	Prediction loss, fairness metric, explainability score → Weighted combination → Governed training objective
Layer 5	Audit reporting and human governance calibration	Human-governed audit and compliance oversight module	Model explanations and fairness outputs → Auditor and policymaker review → Adjusted decision boundaries or retraining

Figure 1.1 presents the role of AI in public policy through five interconnected areas linked to the central node, "AI in Public Policy": agenda-setting, policy formulation, implementation, evaluation, and ethical and governance challenges. In agenda-setting, AI uses NLP to analyze public sentiment, prioritize issues through real-time data, and detect emerging public concerns. In policy formulation, predictive models and machine learning are applied to forecast policy outcomes and scenarios, supporting evidence-based policy design. In implementation, AI uses automation to improve efficiency, deploys tools such as drones and smart systems, and enables real-time feedback for operational adjustments. In evaluation, AI-driven dashboards monitor policy impacts, data analytics track performance, and evidence supports data-based policy adjustments. Ethical and governance challenges include accountable AI practices, privacy and data protection concerns, and the mitigation of bias through fairness and transparency.

AI's role in public policy is transformative, offering policymakers tools that improve the accuracy, efficiency, and inclusiveness of the policy-making process. Yet, the successful implementation of AI in this domain will require careful consideration of ethical challenges and the development of appropriate regulatory frameworks to safeguard public trust and ensure fairness.

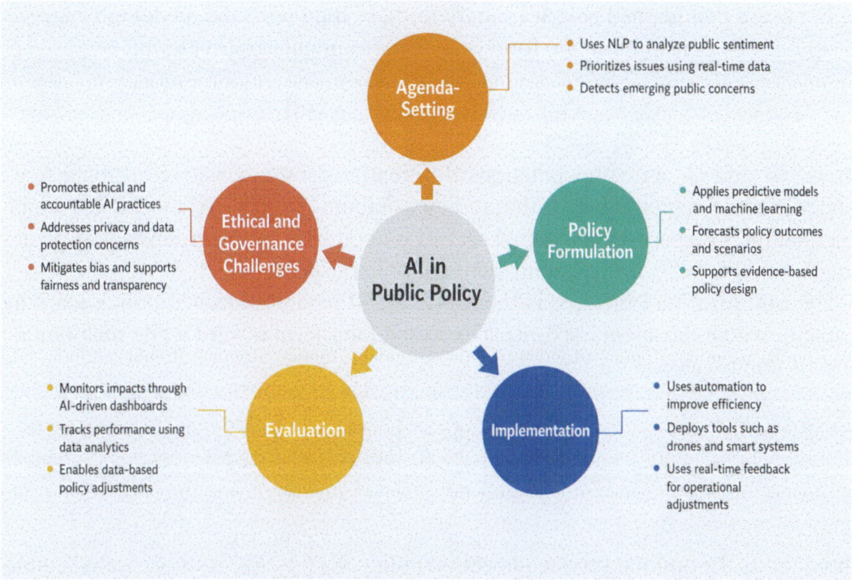

Fig. 1.1 AI in public policy: key areas and applications

1.1.1 Equitable and Efficient Public Services

Artificial intelligence (AI) has emerged as a transformative tool in the delivery of public services, enhancing both operational efficiency and equity. By automating decision-making processes, providing predictive analytics, and optimizing resource allocation, AI is playing an integral role in modernizing public administration, addressing longstanding challenges in service delivery, and ensuring that services are both equitable and efficient (Mergel et al. 2023; Gesk and Leyer 2022).

To operationalize the vision of AI-enhanced public service systems, we introduce a multi-objective deep reinforcement learning (MO-DRL) framework designed to optimize both efficiency and equity in service allocation. This model empowers public administrations to learn policies that adaptively allocate services in response to evolving community needs, budgetary constraints, and fairness goals. The core of this model is the multi-objective reward function, which balances competing outcomes:

$$R_t = \alpha_1 \cdot R_{\text{eff}}(s_t, a_t) + \alpha_2 \cdot R_{\text{equity}}(s_t, a_t)$$

where: R_t is the total reward at time t, $R_{\text{eff}}(s_t, a_t)$ is the efficiency gain from taking action a_t in state s_t, $R_{\text{equity}}(s_t, a_t)$ measures the fairness contribution of the action, and $\alpha_1, \alpha_2 \in \mathbb{R}^+$ are policy weighting coefficients. This equation guides the model to learn decisions that simultaneously improve service productivity and reduce social disparities.

To ensure that learned policies satisfy fairness standards, the model introduces a group fairness constraint in the form of a disparity regularization term:

$$\mathcal{L}_{\text{fair}} = \left|\mathbb{E}_{x\in\mathcal{G}_1}[\hat{y}] - \mathbb{E}_{x\in\mathcal{G}_2}[\hat{y}]\right|$$

where: $\mathcal{G}_1$ and $\mathcal{G}_2$ are subpopulations differentiated by a sensitive attribute (e.g., gender, income, region), and $\hat{y}$ is the predicted outcome from the AI model. This equation quantifies disparity in predicted service outcomes between groups and provides a mathematical basis for enforcing demographic fairness.

The optimization objective is then formulated as a constrained policy learning problem, where the agent maximizes expected long-term reward while maintaining fairness compliance:

$$\pi^* = \arg\max_{\pi} \mathbb{E}_{\pi}\left[\sum_{t=0}^{T} R_t\right] \text{subject to } \mathcal{L}_{\text{fair}} \leq \in$$

where: π^* is the optimal service allocation policy, $\mathbb{E}_{\pi}\left[\sum_{t=0}^{T} R_t\right]$ is the expected cumulative reward over time under policy π, and $\in\in R^+$ is the maximum allowed group disparity. This equation ensures that the model learns policies that are both performance-maximizing and ethically bounded.

This multi-objective deep reinforcement learning model equips public service systems with a powerful tool for making decisions that evolve intelligently over time while maintaining equitable treatment across communities. It integrates performance optimization with normative fairness constraints to ensure that automation supports inclusive governance. By formalizing fairness as part of the learning process and enabling continuous adaptation to public data, the model becomes a cornerstone for building AI-driven governance architectures that are not only efficient but also socially just.

Framework: Multi-Objective Reinforcement Learning Public Service Framework (MO-RL-PSF)

Layered AI Architecture for Equitable and Efficient Public Service Allocation

Layer	Main functionality	AI module(s) & components	Operational scope
Layer 1	Service demand modeling and demographic encoding	Context-aware demand encoding and state representation module	Community needs, budget constraints, population features → Feature preprocessing → State representation vectors
Layer 2	Equity-efficiency reward structuring	Scalarized reward function	Environment state and actions → Efficiency and fairness metrics → Combined reward signal

(continued)

(continued)

Layer	Main functionality	AI module(s) & components	Operational scope
Layer 3	Demographic fairness regularization	Group fairness regularizer	Predicted outcomes by subgroup → Disparity quantification → Penalized loss for bias control
Layer 4	Constrained policy optimization for service allocation	Constrained multi-objective policy learning	Combined reward and fairness threshold → DRL-based optimization → Learned equitable policy π^*
Layer 5	Deployment feedback and ethical adjustment review	Human-in-the-loop ethical oversight and recalibration module	Deployed actions and fairness outcomes → Human review and policy recalibration → Updated learning constraints

One of the key areas where AI enhances public service efficiency is through automation. AI-powered automation systems, such as chatbots and virtual assistants, are now deployed to handle routine tasks that were traditionally handled by human workers. This allows public agencies to streamline administrative processes, providing citizens with 24/7 access to public services without human intervention (Gesk and Leyer 2022). For instance, the use of AI-driven chatbots in municipal services has been shown to drastically reduce response times for citizen inquiries and service requests, thus improving overall satisfaction (Mergel et al. 2023). To enhance the automation of citizen-facing services in public administration, we propose the deployment of a transformer-based natural language understanding (NLU) model optimized with reinforcement learning from human feedback (RLHF). Transformer architectures, particularly those based on bidirectional encoder representations from transformers (BERT) or GPT-based variants, have demonstrated superior capability in capturing language semantics, enabling virtual assistants to understand, process, and respond to user queries with contextual awareness. To fine-tune these systems for optimal real-time responses in civic domains, RLHF allows integration of user satisfaction scores as dynamic feedback, guiding the chatbot's dialogue policy in delivering efficient, contextually accurate, and satisfactory services. The AI-driven model integrating transformer-based NLU with RLHF for real-time public service automation is formulated as:

$$\pi_\theta^* = \arg \underset{\theta}{max} \mathbb{E}_{(x,y,r)}[r \cdot \log\pi_\theta(y|x)]$$

where, π_θ^* denotes the optimal response generation policy of the AI assistant, x represents the citizen's input query, y is the system's response, $\pi_\theta(y|x)$ is the probability assigned to the response y by the transformer model parameterized by θ, and r is the scalar reward reflecting citizen satisfaction or administrative relevance collected via RLHF mechanisms.

This function formalizes the learning objective of the automation system: to adjust its language generation behavior by maximizing expected user-aligned rewards. The transformer's ability to encode context across full input sequences ensures accurate semantic interpretation, while the reward signal allows the AI assistant to refine its linguistic strategies toward clearer, faster, and more useful responses in administrative exchanges. The logarithmic term ensures proper probabilistic scaling, and the reward-weighted expectation incentivizes outputs aligned with public service performance metrics. To effectively update the model via feedback, a policy gradient optimization step is employed using the REINFORCE rule:

$$\nabla_\theta J(\theta) = \mathbb{E}_{(x,y,r)}[r \cdot \nabla_\theta \log \pi_\theta(y|x)]$$

where, $\nabla_\theta J(\theta)$ represents the gradient of the expected reward with respect to model parameters θ, used to improve the policy, and all other terms remain as defined above.

This gradient allows the AI assistant to adjust its internal weights based on the reward received from each response, reinforcing behaviors that yield high citizen satisfaction or resolution accuracy. Over time, the automation system becomes increasingly adept at providing relevant, timely, and context-aware replies across a broad spectrum of administrative services. To ensure interpretability and service consistency, we define the reward function r using a composite feedback mechanism:

$$r = \lambda_1 \cdot f_{\text{sat}}(y) + \lambda_2 \cdot f_{\text{acc}}(y) - \lambda_3 \cdot c(y)$$

where, $f_{\text{sat}}(y)$ is a function capturing citizen satisfaction based on post-response ratings or sentiment scores, $f_{\text{acc}}(y)$ measures factual accuracy or information relevance based on predefined government knowledge bases, $c(y)$ represents the cost in computation or system delay for response y, and λ_1, λ_2, λ_3 are scalar weights tuning the priority of each reward component.

This reward structure ensures that the AI chatbot favors informative, accurate, and low-latency responses while maintaining a high standard of citizen engagement. By incorporating real-world feedback into the optimization loop, the assistant remains adaptive to changing expectations and service standards, continuously refining its capabilities.

This transformer-RLHF framework demonstrates how AI-powered automation can be tailored for public administration needs. It enables real-time, scalable, and accurate interaction with citizens while reducing administrative overhead. The equations ensure that the language generation model learns from feedback loops grounded in user satisfaction and factual correctness, thereby institutionalizing a system of continuous service improvement. As a result, public agencies benefit from enhanced operational efficiency, 24/7 accessibility, and elevated user trust in digital governance platforms.

Framework: Transformer-RLHF Public Service Automation Framework (TRHF-PSAF)
Layered AI Architecture for Real-Time Citizen Interaction and Adaptive Response Optimization

Layer	Main functionality	AI module(s) & components	Operational scope
Layer 1	User query intake and intent representation	Semantic query encoding and context extraction module	Citizen queries → Tokenization, embedding → Semantic context vectors
Layer 2	Policy optimization for contextual response generation	Reward-weighted maximum likelihood policy optimization	Context vectors and historical interactions → Reward-scaled log-likelihood maximization → Optimized response policy π^*
Layer 3	Feedback-driven model refinement	Policy gradient update via reinforce	Citizen feedback signals → Gradient computation → Parameter updates for policy improvement
Layer 4	Reward signal calibration from multi-source metrics	Multi-component reward shaping	Satisfaction, accuracy, and latency scores → Weighted combination → Real-time reward value
Layer 5	Public oversight and conversational auditability	Human-in-the-loop oversight and transparency audit module	Dialogue logs and system responses → Policy analyst review → Trust and consistency checks

Beyond routine automation, AI also plays a crucial role in improving decision-making in public services by offering predictive analytics. Governments now use AI algorithms to analyze vast datasets and predict future service demands. For example, AI can predict peak times for public transportation, helping agencies allocate resources more effectively (Kulal et al. 2024). This predictive capability allows public agencies to preemptively address challenges such as traffic congestion, thereby enhancing service efficiency and improving the quality of life for citizens (Mergel et al. 2023). To anticipate fluctuations in public service demand, particularly in dynamic domains such as urban transit systems, we propose the deployment of a temporal convolutional network (TCN) enhanced with a Bayesian Uncertainty Quantification (BUQ) layer. TCNs are well-suited for predictive analytics due to their ability to model long-range temporal dependencies using dilated causal convolutions, while BUQ augments the model with probabilistic confidence intervals, allowing decision-makers to assess the reliability of predictions in high-stakes scenarios.

The AI-driven predictive model combining TCN with Bayesian Uncertainty Quantification for dynamic public service forecasting is defined by the following formulation:

$$\hat{y}_t = f_{\mathrm{TCN}}(x_{t-w+1:t}; \theta) + \in, \in \sim \mathcal{N}\left(0, \sigma^2\right)$$

where, $\hat{y}_t$ is the predicted public service demand at time t, f_{TCN} represents the output of the Temporal Convolutional Network parameterized by θ, $x_{t-w+1:t}$ is the input sequence of historical demand and contextual features within a temporal window of size w, and $\in$ is the model uncertainty, modeled as a zero-mean Gaussian noise with variance σ^2.

This formulation enables the AI system to learn temporal structures from extensive service datasets, such as transport logs, weather patterns, and event schedules. The convolutional layers capture dependencies across various time lags without relying on recurrence, while the Gaussian error term introduces stochastic variability, allowing the system to generate not only point estimates but also probabilistic forecasts, which are essential for planning under uncertainty.

To update the model in a principled Bayesian manner, the posterior distribution over the weights θ is:

$$\mathcal{L}_{\mathrm{VI}} = \mathbb{E}_{q(\theta)}[\log p(Y|X\ ,\theta)] - D_{\mathrm{KL}}\big[q(\theta)\|p(\theta)\big]$$

where, $\mathcal{L}_{\mathrm{VI}}$ is the variational inference objective, $q(\theta)$ is the approximated posterior over the TCN weights, $p(Y|X\ ,\theta)$ is the likelihood of observed outputs given the input data and model parameters, and D_{KL} is the Kullback–Leibler divergence penalizing deviation from the prior $p(\theta)$.

This loss function ensures that the model not only fits the observed data but also maintains a regularized belief over the parameters. The combination of TCN's temporal precision with Bayesian calibration produces forecasts that are not only highly accurate but also quantified in terms of confidence, offering essential guidance for public resource allocation strategies.

To contextualize the model's forecast in operational decision-making, we define a predictive confidence-aware resource allocation function:

$$\mathcal{A}_t = \arg \underset{a \in \mathcal{U}}{max}\big[\hat{y}_t(a) - \lambda \cdot \sigma_t(a)\big]$$

where, $\mathcal{A}_t$ is the selected public service action (e.g., deploying more buses) at time t, $\hat{y}_t(a)$ is the forecasted demand for action a, $\sigma_t(a)$ is the predictive uncertainty associated with that action, $\mathcal{U}$ is the set of available actions, and λ is a scalar weight representing risk aversion.

This equation allows agencies to prioritize resource allocations not only by expected demand but also by the reliability of the forecast, reducing the likelihood of over- or under-responding to uncertain demand surges. The subtractive risk adjustment ensures robustness, balancing efficiency with caution.

The integration of TCNs with Bayesian Uncertainty Quantification empowers public administrations to move beyond reactive planning toward a predictive governance paradigm. This model equips agencies with precise, uncertainty-aware forecasts of service needs, enabling proactive deployment of infrastructure and personnel. In domains like public transportation, such predictive intelligence improves system resilience, mitigates congestion during peak periods, and enhances overall citizen satisfaction by ensuring services are delivered in anticipation of, rather than in response to, public demand.

Framework: Temporal-Bayesian Public Demand Forecasting Framework (TB-PDFF)
Layered AI Architecture for Predictive Service Allocation Under Uncertainty

Layer	Main functionality	AI module(s) & components	Operational scope
Layer 1	Temporal data aggregation and preprocessing	Temporal data encoding and sequence structuring module	Historical service logs, weather, and events → Normalization and windowing → Sequence data matrix
Layer 2	Demand forecasting via sequence modeling	Dilated temporal convolutional forecasting	Input sequence of temporal features → Convolutional inference → Predicted future demand
Layer 3	Uncertainty estimation for forecast reliability	Variational Bayesian neural network	Predicted outputs and model weights → Posterior inference → Confidence interval generation
Layer 4	Decision-making under forecast risk	Risk-adjusted decision rule	Demand predictions and uncertainty scores → Risk-weighted utility scoring → Selected service response
Layer 5	Administrator oversight and strategic review	Human-in-the-loop strategic oversight and policy refinement module	Forecast reports and action logs → Policy validation → Iterative strategy refinement

Another significant advantage of AI in public service delivery is its capacity to ensure equity in access and service quality. By analyzing demographic data, AI systems can help identify underserved communities and ensure that public services are distributed more equitably. This is particularly important in sectors such as healthcare, where AI can assist in targeting healthcare resources to vulnerable populations (Ruijer et al. 2022). AI-driven data analysis can highlight patterns of service inequality, allowing public administrators to implement targeted interventions that address these disparities (Ruijer et al. 2022). To operationalize fairness in public service distribution, we introduce a fairness-constrained resource allocation model built upon extreme gradient boosting (XGBoost) for prediction accuracy, combined with a Fairness Constraints Optimization Layer to ensure demographic equity in outcomes. XGBoost is selected for its robust handling of heterogeneous and high-dimensional data, such as population demographics, service history, and socioeconomic indicators. To integrate equity into the prediction-to-decision pipeline, fairness constraints are embedded into the optimization objective, ensuring that predictions do not amplify existing disparities across sensitive groups.

The AI-driven model integrating XGBoost with fairness-aware constrained optimization for equitable public service targeting is expressed as:

$$\min_{\mathrm{w}}\left[\sum_{i=1}^{n}\ell(f_{\mathrm{w}}(x_i), y_i) + \lambda\cdot\Omega(f_{\mathrm{w}})\right]\text{subject to }\left|\mu_{g_1}-\mu_{g_2}\right|\leq\in$$

where, $f_{\mathrm{w}}(x_i)$ is the prediction function of the XGBoost model for input feature vector x_i with parameters w, y_i is the true service outcome, $\ell(\cdot)$ is the loss function (e.g., squared error), $\Omega(f_{\mathrm{w}})$ is the regularization term controlling model complexity, λ is the regularization weight, μ_{g_1} and μ_{g_2} represent average prediction outcomes for two different demographic groups (e.g., privileged vs. underserved), and $\in$ is a fairness threshold bounding the permissible disparity between groups.

This formulation enables the model to learn highly accurate predictions of service need while simultaneously enforcing demographic parity, ensuring that no population subgroup is consistently disadvantaged by the model's decisions. The constraint term prevents large disparities in predicted service delivery across demographic dimensions such as income, age, race, or geographic location.

To reinforce equity at the decision-making level, we define a demographic-aware utility optimization function that allocates public resources proportionally based on predicted need and adjusted fairness weights:

$$\mathcal{R}_j = \frac{f_{\mathrm{w}}\left(x_j\right)}{1 + \gamma \cdot D\left(x_j\right)}$$

where, $\mathcal{R}_j$ is the resource allocation score for community or individual $j, f_{\mathrm{w}}\left(x_j\right)$ is the predicted need score from the XGBoost model, $D\left(x_j\right)$ is a demographic disadvantage index capturing social vulnerability (e.g., deprivation index, healthcare access deficit), and γ is a hyperparameter controlling the extent of fairness correction.

This function ensures that individuals or communities with higher disadvantage receive proportionally more attention, even when raw predictive need may be similar to others. The denominator down-weights allocations in already well-served regions, thereby counterbalancing structural inequalities and reinforcing inclusive service planning.

To support monitoring and transparency, we define the group-level equity evaluation function as:

$$\Delta_{\mathrm{eq}} = \underset{g_1, g_2}{max} \left| \frac{1}{\left|G_{g_1}\right|} \sum_{j \in G_{g_1}} \mathcal{R}_j - \frac{1}{\left|G_{g_2}\right|} \sum_{j \in G_{g_2}} \mathcal{R}_j \right|$$

where, Δ_{eq} quantifies the largest discrepancy in average allocated resources across any two demographic groups g_1 and g_2, and G_{g_k} is the population set of group g_k.

This equity metric is essential for assessing the fairness of implemented interventions, providing public administrators with a quantitative indicator of disparity that guides corrective policies and resource redistribution over time.

This fairness-constrained XGBoost framework transforms raw demographic and service data into equitable decision-support tools. By embedding fairness directly into the predictive and allocative phases, the model avoids perpetuating historical biases and ensures that public services especially in critical areas such as healthcare are directed toward communities in need. The equations collectively advance a proactive and just approach to governance, reinforcing the principle that equity must be measured, modeled, and enforced through deliberate algorithmic design.

Framework: Fairness-Constrained XGBoost Public Service Allocation Framework (FC-XGBoost-PSA)

Layered AI Architecture for Equitable and Demographic-Aware Service Distribution

Layer	Main functionality	AI module(s) & components	Operational scope
Layer 1	Demographic feature encoding and preprocessing	Demographic data integration and feature structuring module	Raw population, service, and geographic data → Normalization and feature engineering → Model-ready structured dataset
Layer 2	Fairness-constrained prediction training	Fairness-constrained XGBoost training	Input features and outcome labels → Boosted tree model training → Accuracy-fairness balanced predictions
Layer 3	Equity-informed allocation scoring	Demographic-aware utility function	Predicted service needs and disadvantage index → Score computation → Adjusted allocation priorities
Layer 4	Equity disparity evaluation	Group equity evaluation metric	Allocation scores by subgroup → Disparity measurement → Group fairness quantification
Layer 5	Policy oversight and redistribution calibration	Human-in-the-loop equity oversight and policy calibration module	Equity evaluation reports → Human feedback and policy update → Adjusted intervention strategy

However, while AI has the potential to revolutionize public service delivery, it is not without challenges. Ethical concerns around data privacy, transparency, and accountability have been raised, particularly with regard to automated decision-making systems. The use of AI in public administration raises questions about fairness, particularly when algorithms are used to determine eligibility for services such as social welfare or healthcare (Mergel et al. 2023). To mitigate these risks, public sector organizations need to develop strong governance frameworks that ensure the responsible use of AI, with an emphasis on transparency, accountability, and the protection of citizens' rights (Mergel et al. 2023). To address the critical challenges associated with fairness, transparency, and rights protection in AI-based public decision-making, we introduce a Fair Adversarial Neural Network (FANN). This model integrates fairness constraints into learning objectives while incorporating an auditable latent representation, ensuring decisions made by public AI systems are both equitable and inspectable. The core of the framework is the task-specific prediction function (e.g., eligibility classification):

$$\hat{y} = f_\theta(x)$$

where: $\hat{y}$ is the AI system's decision (e.g., service eligibility), f_θ is the neural network model parameterized by weights θ, and x is the input vector representing citizen features (e.g., income, age, health status). This equation produces a decision that will later be regularized for fairness and transparency.

To ensure fairness, we implement an adversarial debiasing function, where an adversary attempts to predict sensitive attributes from the model's internal representation, while the main model tries to minimize this leakage:

$$\mathcal{L}_{\text{adv}} = \mathbb{E}_{x \sim \mathcal{D}}\left[\ell\left(g_\phi(f_\theta(x)), s\right)\right]$$

where: ℓ is the loss function, g_ϕ is the adversarial predictor parameterized by ϕ, s is the sensitive attribute (e.g., race, gender), and $f_\theta(x)$ is the latent representation used for the decision. This adversarial term penalizes the model for encoding demographic attributes that may cause biased outcomes.

The final training objective becomes a min–max optimization balancing predictive accuracy and fairness:

$$\underset{\theta}{min}\,\underset{\phi}{max}\left[\mathcal{L}_{\text{task}}(f_\theta(x), y) - \lambda \cdot \mathcal{L}_{\text{adv}}\left(g_\phi(f_\theta(x)), s\right)\right]$$

where: $\mathcal{L}_{\text{task}}$ is the primary loss (e.g., cross-entropy for classification), $\mathcal{L}_{\text{adv}}$ is the adversarial fairness loss, and λ is a scalar hyperparameter that controls the fairness–accuracy trade-off. This formulation explicitly trains the system to maximize fairness without severely compromising performance.

To ensure transparency and auditability, we introduce a traceable explanation layer, using SHAP-based decomposition:

$$\hat{y} = \sum_{k=1}^{K} \phi_k x_k$$

where: ϕ_k is the SHAP value of feature x_k, indicating its contribution to the output $\hat{y}$. This formulation allows auditors and stakeholders to understand and verify the basis of each AI-driven decision.

The fair adversarial neural network model directly addresses core governance requirements for AI in the public sector. It enables agencies to make high-stakes decisions such as eligibility for healthcare or welfare with explicit mathematical safeguards for fairness, explainability, and sensitivity isolation. By ensuring that sensitive attributes cannot be reconstructed from latent representations and offering a traceable explanation mechanism, this model operationalizes public values such as justice, equity, and accountability. It provides a computational foundation for a rights-respecting AI infrastructure, ensuring that technological innovation in government remains ethically anchored and socially responsible.

Framework: Fair Adversarial Neural Network for Ethical Public Decision-Making (FANN-EPDM)

Layered AI Architecture for Fairness, Transparency, and Rights Protection in Public Services

Layer	Main functionality	AI module(s) & components	Operational scope
Layer 1	Citizen feature encoding and preprocessing	Citizen data preparation and feature encoding module	Raw citizen data (income, age, health) → Feature encoding → Structured input vector
Layer 2	Eligibility classification prediction	Neural eligibility classifier	Encoded input features → Feedforward neural network → Initial service decision output
Layer 3	Demographic leakage detection and removal	Sensitive attribute adversarial filter	Latent representations → Adversarial training loop → Debiased internal encoding
Layer 4	Fairness-constrained training optimization	Adversarial fairness training objective	Task loss and adversarial loss → Min–max balancing → Trade-off optimization of fairness vs. performance
Layer 5	Explainability and decision justification	SHAP-based attribution explanation	Trained model and citizen input → Attribution decomposition → Transparent decision rationale
Layer 6	Ethical oversight and compliance reporting	Human-in-the-loop ethical governance and compliance module	Explanatory logs and fairness scores → Review by policymakers → Corrective actions and ethical compliance

Figure 1.2 illustrates the role of AI in enhancing public services through four main areas: efficiency, decision-making, equity, and ethics. AI improves efficiency by automating tasks, such as using chatbots for 24/7 service, leading to faster response times. In decision-making, AI employs predictive analytics to anticipate demand, like public transportation peaks, and optimize resources to address issues like traffic. AI promotes equity by identifying underserved areas and providing targeted support, such as healthcare for vulnerable groups. Ethical challenges include concerns about privacy, fairness, accountability, and the need for transparency in AI use within public services. Each area is color-coded and connected to the central theme.

AI offers immense potential to enhance the efficiency and equity of public service delivery. By automating routine tasks, offering predictive insights, and addressing service disparities, AI is reshaping the landscape of public administration. However, public organizations must address the ethical challenges associated with AI to ensure that its benefits are realized in an equitable and fair manner.

Case Studies

The use of artificial intelligence (AI) in public policy has expanded significantly in recent years, demonstrating how AI technologies can enhance the effectiveness,

Fig. 1.2 AI in equitable and efficient public services

efficiency, and fairness of public administration. Several case studies from various sectors provide insights into the successful implementations of AI in public policy, highlighting its potential for automating decision-making, improving operational processes, and fostering public trust.

U.S. Federal Agencies and AI Integration

A comprehensive study by Engstrom et al. (2020) explores the use of AI in multiple U.S. federal agencies, demonstrating how AI has been successfully deployed in regulatory enforcement, public engagement, and adjudication processes. Agencies such as the Securities and Exchange Commission (SEC) have used machine learning algorithms to predict enforcement actions, while the Customs and Border Protection (CBP) has employed AI for facial recognition in border security. Another notable case is the Social Security Administration's use of AI to correct adjudicatory errors in disability benefit claims (Engstrom et al. 2020). These cases underscore how AI is not only automating routine processes but also enhancing transparency and reducing human error in decision-making.

Flanders Investment and Trade (FIT)

FIT, a trade promotion organization based in Flanders, Belgium, adopted AI to improve its customer service operations. The organization implemented an AI-driven question-answering algorithm to automate responses to routine inquiries from Flemish companies seeking to expand internationally (Nurski 2023). By using natural language processing, the AI system retrieved answers from a large dataset of previously answered questions, significantly reducing the workload of human advisors and allowing them to focus on more complex inquiries. The adoption process involved a careful alignment of AI technology with existing business processes, highlighting the importance of integrating human-centered AI systems that support, rather than replace, human workers (Nurski 2023).

AI Watch by the European Commission

The European Commission's AI Watch initiative provides an extensive overview of AI implementations across Europe, detailing over 230 AI use cases in public services. This initiative tracks the progress of AI beyond pilot projects, focusing on sustainable implementations in sectors such as healthcare, education, and transportation. In one example, AI was used to predict traffic congestion and optimize public transportation schedules in several European cities, improving resource allocation and reducing delays (European Commission 2021). The AI Watch report emphasizes the need for a balanced approach between AI promotion and regulation to ensure that AI technologies are adopted responsibly in public governance (European Commission 2021).

Challenges and Opportunities

While these case studies highlight the success of AI implementations in public policy, they also underscore significant challenges. Data quality, algorithmic transparency, and accountability remain critical concerns. For instance, Engstrom et al. (2020) point out that the lack of explainability in some AI models can hinder public trust, especially in high-stakes decision-making contexts such as social welfare distribution. Similarly, the European Commission emphasizes the importance of ensuring that AI systems are transparent and aligned with ethical standards, particularly in areas involving personal data and public engagement (European Commission 2021).

The successful integration of AI in public policy depends on addressing these challenges through strong governance frameworks and cross-sector collaboration. By fostering an environment where AI can be developed and implemented responsibly, public agencies can fully realize the potential of AI to enhance service delivery, increase efficiency, and promote fairness in public administration.

1.2 The Risk of Technological Unemployment

The risk of technological unemployment posed by AI and automation has been a major concern for both scholars and policymakers. The potential for AI to disrupt labor markets lies in its ability to automate tasks that were once the sole domain of human workers, resulting in a displacement effect. However, this displacement is counterbalanced by productivity gains and the creation of new tasks, which could mitigate the negative impact on overall employment.

1.2.1 Displacement Effect and Labor Market Disruption

At the core of the labor market disruption caused by AI is the displacement effect, where AI and robotics substitute labor in tasks that humans previously performed. This substitution can lead to a decrease in the demand for labor in those specific tasks, particularly in routine, low- and medium-skilled jobs. Acemoglu and Restrepo (2020) highlight how this process of automation is fundamentally reshaping industries. In sectors like manufacturing, clerical work, and even healthcare, machines are taking over repetitive tasks, leaving human workers redundant (Acemoglu and Restrepo 2020). For example, routine manual jobs in production and office administration have been particularly vulnerable to AI-driven automation (Acemoglu and Restrepo 2020). To mathematically capture the displacement effect and its cascading consequences across labor sectors, we propose a dynamic sectoral reinforcement learning model (DSRLM), integrating reinforcement learning (RL) for AI-driven substitution behavior and elasticity of substitution (σ) for labor-technology tradeoff quantification. This model simulates how an AI agent learns to replace human labor where automation yields higher utility and lower marginal cost, while the economy adapts through wage adjustments and labor migration.

The AI-driven DSRLM for modeling labor displacement through sectoral substitution is formally expressed as:

$$V(s_t) = \underset{a_t \in \mathcal{A}}{max}\left[R(s_t, a_t) + \gamma \cdot \mathbb{E}_{s_{t+1}}\left[V(s_{t+1})\right]\right]$$

where, $V(s_t)$ is the value function representing the expected long-term utility of substituting human labor with AI in state s_t, $a_t \in \mathcal{A}$ denotes the automation action (i.e., substituting a job task), $R(s_t, a_t)$ is the immediate reward from automation (cost savings, productivity gain), $\gamma \in (0, 1)$ is the discount factor, and $\mathbb{E}_{s_{t+1}}\left[V(s_{t+1})\right]$ is the expected future utility in state s_{t+1}.

This reinforcement learning structure captures the strategic behavior of firms or sectors adopting automation over time. The agent receives short-term gains from automation while learning the long-term implications on productivity, labor costs, and regulatory constraints. The recursive Bellman equation allows simulation of automation strategies across different economic states.

The immediate reward function $R(s_t, a_t)$, capturing both economic and labor displacement factors, is defined as:

$$R(s_t, a_t) = \alpha \cdot \Delta P_t(a_t) - \beta \cdot \Delta L_t(a_t)$$

where, $\Delta P_t(a_t)$ is the productivity gain resulting from the automation of task a_t, $\Delta L_t(a_t)$ is the corresponding reduction in labor demand or job displacement, and α, β are scaling parameters representing the priority weighting between productivity and labor impact. This function quantifies how each automation action affects the labor market: gains in productivity yield economic returns, while job losses generate a social cost. Balancing these outcomes helps policymakers simulate interventions such as taxation of automation or upskilling subsidies.

To capture the technological substitution elasticity between labor and AI capital in a given sector, we define the AI-labor substitution production function as:

$$Y_t = \left[\phi \cdot L_t^{\frac{\sigma-1}{\sigma}} + (1-\phi) \cdot K_t^{\frac{\sigma-1}{\sigma}}\right]^{\frac{\sigma}{\sigma-1}}$$

where, Y_t is the total sectoral output at time t, L_t is the labor input, K_t is the AI capital input (i.e., machines or algorithms performing automated tasks), ϕ is the labor share parameter, and σ is the elasticity of substitution, governing how easily AI replaces human labor.

This constant elasticity of substitution (CES) function models how industries reconfigure their production processes in response to automation pressures. A higher σ implies easier substitution and greater displacement risk for human workers.

This reinforcement learning-based framework enriched with elasticity-driven production theory enables a dynamic understanding of AI-induced labor market disruption. By simulating task-level automation decisions and quantifying trade-offs between productivity and job displacement, this model provides both analytical clarity and practical insights into structural shifts within the labor economy. The RL reward structure mirrors real-world employer strategies in maximizing efficiency, while the CES production function underscores the substitutability threshold beyond which routine jobs face systemic elimination. This integrated AI-economic model equips researchers and policymakers with a predictive tool to design inclusive transition strategies, reskilling programs, and AI governance policies to mitigate the displacement effect in a rapidly evolving labor market.

Framework: Dynamic Sectoral Reinforcement Learning Model (DSRLM)
Layered AI Architecture for Labor Displacement and Automation Substitution Modeling

Layer	Main functionality	AI module(s) & components	Operational scope
Layer 1	Automation decision optimization via RL	Reinforcement learning (RL)	Labor state information → Selection of automation actions → Refined automation strategies
Layer 2	Reward evaluation incorporating productivity and displacement	Custom multi-objective reward function	Automation outcomes → Trade-off computation → Net reward estimation
Layer 3	Sectoral output modeling via AI-labor substitution	Constant elasticity of substitution (CES) function	Labor and AI resource levels → Substitution modeling → Sectoral performance output
Layer 4	Policy simulation for reskilling or automation taxation	Policy impact simulation and scenario analysis module	Displacement and reward trends → Policy impact evaluation → Strategic planning insights

(continued)

(continued)

Layer	Main functionality	AI module(s) & components	Operational scope
Layer 5	Monitoring of displacement dynamics and migration	Labor dynamics monitoring and adaptive feedback module	Labor shifts and automation rates → Aggregated displacement tracking → Monitoring reports and feedback

While artificial intelligence offers substantial efficiency gains and automation potential, its large-scale deployment has also introduced a systemic decoupling of productivity and wage growth, particularly across labor markets performing routine or semi-skilled tasks. According to findings from the National Bureau of Economic Research, automation increases per-worker output but does not yield commensurate gains in real wages, thereby contributing to widening income inequality (Acemoglu and Restrepo 2020).

This phenomenon commonly referred to as the displacement effect reflects a broader structural dynamic in which AI and automation technologies augment high-skill labor while substituting routine tasks, a process consistent with task polarization theory (Autor 2015). Autor demonstrates that automation and computerization systematically replace routine, codifiable tasks such as bookkeeping, clerical functions, and repetitive production work, while simultaneously enhancing high-skill abstract activities including problem-solving, managerial decision-making, and creative innovation. At the same time, technological change supports the expansion of low-skill manual service occupations that are less easily automated. The combined outcome is a polarized labor market characterized by employment growth at the upper and lower ends of the skills and wage distribution, accompanied by a marked erosion of middle-skill routine occupations.

As AI continues to reconfigure the value of labor inputs, the labor share of national income—the proportion of GDP accruing to workers—has exhibited a measurable decline. Korinek et al. (2021) underscore this trend, emphasizing how AI exacerbates capital-biased inequality by concentrating economic returns in the hands of technology owners and highly skilled professionals.

These dynamics pose significant challenges for public governance and inclusive development. Left unaddressed, they risk entrenching systemic inequalities that compromise social cohesion and trust in public institutions. It becomes imperative, therefore, that AI policy frameworks include redistributive mechanisms, such as adaptive taxation, universal reskilling programs, and equitable access to AI-generated productivity dividends. Such mechanisms are essential to prevent what could otherwise become a technologically enabled form of structural exclusion.

In this context, the role of governments is not only to adopt AI but to institutionalize its ethical deployment, ensuring that the benefits of innovation extend across socio-economic strata. This requires integrating macroeconomic foresight into AI governance architectures, aligning technological progress with social justice and democratic sustainability.

1.2.2 Countervailing Forces: Productivity Gains and Task Creation

Despite these negative effects, automation also generates a productivity effect. AI-driven automation reduces production costs, making goods and services cheaper and increasing demand in non-automated sectors. This productivity boost can lead to the creation of new labor-intensive tasks, particularly in industries where human interaction or creativity is paramount (Babina et al. 2021). For example, in AI-driven industries, companies require a highly educated workforce with STEM expertise, as investments in AI tend to increase the demand for workers with technical skills (Babina et al. 2021). This shift has been observed across various sectors, as firms adopting AI tend to hire more workers in technical roles while decreasing their reliance on middle-management and routine roles (Babina et al. 2021). To quantify the emergence of new tasks and the associated skill demand driven by AI-induced productivity gains, we develop a dynamic task-creation model with productivity-driven expansion, embedded in a multi-class skill demand function. This model captures how automation reallocates economic surplus toward new labor-intensive functions requiring high-skill capabilities, particularly in technical fields such as data science, engineering, and system design.

The AI-driven model integrating productivity expansion with task-induced skill reallocation is formulated as follows:

$$\Delta H_t = \eta \cdot \frac{\partial Y_t}{\partial K_t} \cdot \frac{\partial T_{\text{new}}}{\partial H}$$

where, ΔH_t is the net increase in demand for high-skilled human capital at time t, $\frac{\partial Y_t}{\partial K_t}$ is the marginal productivity of AI capital (reflecting production gains from automation), $\frac{\partial T_{\text{new}}}{\partial H}$ is the marginal responsiveness of new task creation to the availability of high-skilled labor, and η is a scale parameter linking productivity growth to job creation intensity.

This formulation demonstrates how increased efficiency from AI capital not only enhances output but also generates structural opportunities for new tasks provided that sufficient human capital exists to absorb and operationalize them. The final term ensures that task creation is bounded by the availability of educated, specialized workers (e.g., STEM professionals), reflecting real-world skill bottlenecks.

To track the demand evolution across skill categories, we define a multi-class skill demand expansion function as:

$$D_t^{(i)} = \omega_i \cdot T_{\text{new}}(K_t, A_t) \cdot \chi_i$$

where, $D_t^{(i)}$ is the demand for skill class i (e.g., data scientists, system architects, engineers) at time t, ω_i is the sector-specific intensity weight for skill class i, $T_{\text{new}}(K_t, A_t)$ denotes the number of new tasks created as a function of AI capital K_t and total automation advancement level A_t, and χ_i is the relevance index of skill class ii to the newly emerging task domain.

This equation explains how the new economic structure reallocates labor demand toward specific technical profiles. As AI-driven productivity grows, firms require complementary human expertise to deploy, manage, and refine intelligent systems, particularly those functions involving creativity, design, reasoning, and ethical oversight.

We also define a Task-Creation Elasticity Function to model how responsive task formation is to productivity increases:

$$\in_{T,Y} = \frac{\partial T_{\text{new}} / T_{\text{new}}}{\partial Y_t / Y_t}$$

where, $\in_{T,Y}$ is the elasticity of task creation with respect to total output growth. A higher $\in_{T,Y}$ indicates that as output increases due to automation and efficiency, there is a proportionally larger emergence of new roles requiring human creativity or technical depth.

This dynamic AI-driven framework mathematically formalizes the counterbalancing mechanism by which automation, despite displacing routine labor, simultaneously induces demand for highly skilled human workers through new task creation. The productivity-function-driven increase in task formation ensures that AI is not merely a substitute for labor but also a catalyst for structural labor reallocation. The inclusion of elasticity and skill-demand mappings provides a quantitative basis for anticipating workforce shifts in the age of intelligent automation. As Babina et al. (2021) observed, firms adopting AI technologies do not uniformly eliminate jobs; rather, they restructure their labor force toward creative, adaptive, and technically proficient roles, reshaping the human capital landscape of future economies.

Framework: Productivity-Driven Task Creation and Skill Demand Model (PTC-SDM)
Layered AI Architecture for Modeling Productivity Gains and High-Skill Labor Reallocation

Layer	Main functionality	AI module(s) & components	Operational scope
Layer 1	High-skill demand from productivity expansion	Productivity-to-task gradient model	Automation productivity growth → Gradient transformation → High-skill demand projection
Layer 2	skill-class specific labor allocation	Multi-class skill allocation mapping	New tasks from AI systems → Sector-wise skill matching → Skill-specific labor demand profiles
Layer 3	elasticity estimation of task creation	Elasticity of task creation to output	Output growth indicators → Elasticity computation → Task formation sensitivity estimation
Layer 4	Scenario analysis of AI workforce restructuring	Workforce transition simulation and skill dynamics module	AI adoption trends → Simulation of skill transitions → Strategic skill development forecasting

(continued)

(continued)

Layer	Main functionality	AI module(s) & components	Operational scope
Layer 5	Policy insight for STEM investment	Human-capital policy optimization and investment planning module	Skill bottlenecks and elasticity outcomes → STEM education and training gap analysis → Policy feedback mechanisms

In addition to displacing routine labor, artificial intelligence also serves as a platform for new forms of human employment, a dynamic captured in labor economics as the reinstatement effect. As defined by Acemoglu and Restrepo (2020), this phenomenon refers to the creation of novel occupations in response to the deployment of AI technologies. These tasks ranging from AI training, model supervision, ethical oversight, to the development of creative and emotionally intelligent content fall outside the current scope of machine capabilities and thus depend inherently on human input.

The emergence of such roles underscores a broader structural shift in labor demand from task execution to task design and oversight, reflecting a transition from manual and routine functions to abstract, cognitive, and affective labor. This shift is not merely technological but institutional, requiring education systems, labor markets, and industrial policy to co-evolve alongside AI capabilities. Unlike traditional automation, which substitutes existing roles, AI's reinstatement effect reconfigures the division of labor delegating mechanical precision to machines while elevating human functions related to adaptability, judgment, and creativity.

Recognizing and supporting the reinstatement effect is essential for crafting inclusive AI policies. It justifies investment in human-centered upskilling initiatives and calls for interdisciplinary education that integrates technical, ethical, and creative proficiencies. In this light, AI is not simply a driver of labor displacement, but a catalyst for occupational innovation enabling the design of future work ecosystems in which human agency remains central.

1.2.3 The Skills Gap and Labor Market Adjustment

The labor market's ability to adjust to these changes is crucial for mitigating technological unemployment. However, the rapid pace of AI adoption has led to a skills gap in the workforce. Many workers do not possess the necessary skills to transition into new roles created by AI technologies. This mismatch between the supply of labor and the demand for AI-specific skills slows down the labor market's ability to adjust to technological advancements (Acemoglu and Restrepo 2020). For example, while demand for high-skilled workers, particularly those with STEM backgrounds, has increased, workers in routine jobs, such as manufacturing or office work, are less able to transition into AI-driven sectors (Korinek et al. 2021). To mathematically capture how the skills gap impedes labor market adjustment in response to AI adoption, we formulate a skills adjustment friction model (SAFM), integrating an AI-specific skill mismatch index into a dynamic transition probability function. This model quantifies the probability of successful labor reallocation as a function of skill alignment, retraining efficiency, and sectoral absorptive capacity.

The AI-driven SAFM for modeling skill-based labor market rigidity is expressed as:

$$P_{\text{trans}}(i,t) = \psi_i \cdot \exp(-\kappa \cdot M_i(t))$$

where, $P_{\text{trans}}(i,t)$ is the probability that a worker from occupation i successfully transitions into an AI-related role at time t, ψ_i is the baseline mobility coefficient for occupation i (capturing historical adaptability), $M_i(t)$ is the skill mismatch index at time t, and κ is the rigidity sensitivity parameter quantifying how severely mismatch reduces transition probability. This exponential decay formulation reflects that higher mismatch significantly reduces transition chances. The term $M_i(t)$ aggregates differences between existing worker skill profiles and the evolving skill requirements of AI-intensive industries.

To define this central element, we introduce the skill mismatch index as:

$$M_i(t) = \sum_{k=1}^{K} \left| s_{i,k}(t) - d_k(t) \right|$$

where, $s_{i,k}(t)$ is the average level of skill κ (e.g., programming, data analysis, machine learning) present in workers from occupation i, $d_k(t)$ is the skill demand in the target AI domain for skill κ, and K is the total number of skill dimensions considered. This index captures the absolute deviation between the skill supply and demand vectors. Larger cumulative differences across all skills imply that retraining would need to be more extensive and that frictional unemployment may be prolonged.

To represent the impact of retraining policies, we define a retraining effectiveness function that modifies skill supply trajectories:

$$s_{i,k}(t+1) = s_{i,k}(t) + \delta \cdot R_{i,k}(t)$$

where, $s_{i,k}(t+1)$ is the updated skill level in dimension κ after retraining at time $t+1$, $R_{i,k}(t)$ is the retraining resource allocated to occupation i in skill k, and δ is the training absorption coefficient that translates investments into effective skill gains. This dynamic update allows policymakers to model the responsiveness of the labor force to targeted reskilling programs. It creates a pathway for reducing mismatch $M_i(t)$ over time, thereby raising $P_{\text{trans}}(i,t)$ and facilitating labor market equilibrium.

Finally, we define a macro-adjustment indicator summarizing the system-wide labor flexibility in response to AI transition:

$$\Theta_t = \frac{1}{N} \sum_{i=1}^{N} P_{\text{trans}}(i,t)$$

where, Θ_t is the average transition probability across all occupations $i \in \{1, ..., N\}$ at time t. This indicator provides a single interpretable metric that captures the overall readiness of the labor market to adjust to structural changes induced by AI technologies.

This skill mismatch and transition modeling framework introduces a robust AI-driven methodology to quantify the bottlenecks that slow labor market adaptation to technological change. By embedding skill supply–demand divergence directly into transition probabilities, and by explicitly modeling retraining impacts, the SAFM highlights how targeted interventions can transform labor rigidity into adaptability.

Framework: Skills Adjustment Friction Model (SAFM)

Layered AI Architecture for Modeling Labor Rigidity and Skill Transition Under AI

Layer	Main functionality	AI module(s) & components	Operational scope
Layer 1	Transition probability modeling with skill friction	Transition probability decay model	Skill mismatch index → Exponential transformation → Labor mobility probability
Layer 2	Mismatch index calculation	Skill mismatch quantification	Worker skill profiles and AI skill demand → Absolute deviation computation → Mismatch index value
Layer 3	Reskilling impact projection	Reskilling dynamics model	Training resources allocation → Skill level updates → Adjusted skill supply trajectory
Layer 4	Macro readiness estimation	Macro-adaptability indicator	Occupation-level transition probabilities → Aggregation → System-wide adaptability score
Layer 5	Policy planning for retraining investment	Human-capital investment and retraining strategy module	Mismatch trends and adjustment elasticity → Intervention scenario analysis → Strategic planning output

While the SAFM quantifies how skill mismatches hinder labor market transitions, it is critical to recognize that effective AI integration depends as much on human capital infrastructure as on technological deployment. Firms investing in AI must complement these advancements with parallel investments in workforce development, creating ecosystems in which machine and human intelligence evolve in tandem (Babina et al. 2021).

This dual transformation demands targeted upskilling initiatives within companies and the recalibration of public education systems to meet new labor demands. As Korinek et al. (2021) argue, without proactive interventions, the displacement effects of automation will deepen existing inequalities, particularly in communities where access to technical education is limited. Consequently, governments must play a central role in bridging the AI skills gap by expanding STEM education pipelines, funding reskilling programs, and incentivizing private sector training.

Moreover, a modern workforce strategy must emphasize not only technical competencies but also cross-disciplinary agility preparing individuals for hybrid roles that combine data literacy, design thinking, ethics, and human-centered innovation. Without these reforms, the structural bottlenecks modeled in SAFM will persist, and the benefits of AI will be captured by a narrowing segment of the population.

Figure 1.3 depicts the risk of technological unemployment associated with AI and automation through three main areas: displacement effect and labor market disruption, countervailing forces, and skills gap and labor market adjustment. The displacement effect shows how AI and robotics can substitute human labor in routine tasks, making low- and medium-skilled jobs, such as manufacturing and clerical work, especially vulnerable, and potentially contributing to wage stagnation and income inequality as demand falls for certain jobs. Countervailing forces show that productivity gains may lower costs and create demand in other sectors, increase demand for high-skilled technical roles, particularly STEM jobs, and generate new roles such as AI training and machine supervision through the reinstatement effect. The skills gap section highlights the mismatch between existing workforce skills and AI-related job demands, emphasizing education and training in STEM and technical skills, as well as the role of governments and companies in reskilling and upskilling initiatives. The figure therefore presents both the employment risks created by AI-driven automation and the adaptive mechanisms needed to respond to them.

The potential for AI to disrupt labor markets and displace workers is significant, particularly in routine jobs that are highly susceptible to automation. The displacement effect, driven by AI's ability to replace labor in certain tasks, poses a direct threat to job security in low- and medium-skilled sectors. However, this risk is mitigated by productivity gains and the creation of new tasks that AI generates, particularly in industries that require technical expertise. To fully realize the benefits of AI while mitigating the risks of technological unemployment, it is essential to address the skills gap and provide retraining opportunities for workers. In this way, the transition to an AI-driven economy can be managed in a way that fosters inclusive growth and minimizes inequality.

1.3 Strategies for Workforce Transition and Reskilling

The integration of artificial intelligence (AI) into industries has raised significant concerns about workforce displacement, particularly the risk of technological unemployment. This transition, however, also highlights the urgent need for workforce transition strategies and reskilling programs to address the evolving demands of an AI-driven economy.

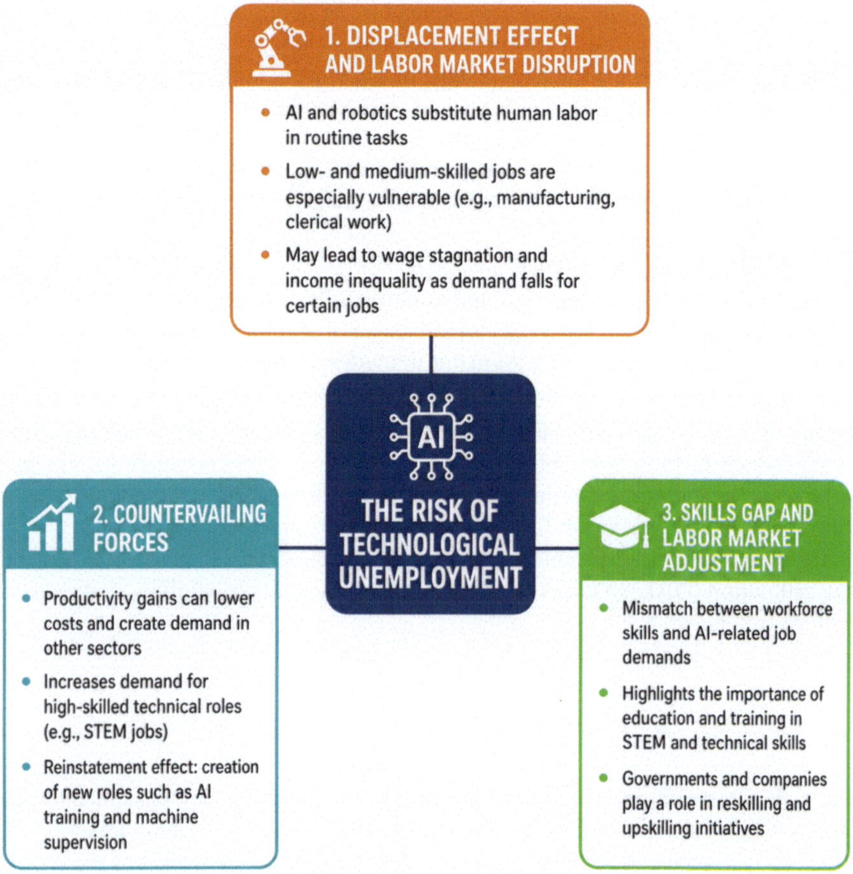

Fig. 1.3 The risk of technological unemployment

1.3.1 The Imperative of Reskilling and Upskilling

AI and automation are transforming industries by replacing routine tasks with machine-driven processes. As a result, traditional job roles are diminishing, while new roles requiring advanced technical and cognitive skills are emerging. The primary challenge is to bridge the skills gap by equipping the current workforce with the necessary capabilities to thrive in this new landscape (Bobitan et al. 2024). The World Economic Forum (2020) has projected that nearly half of the global workforce will need significant reskilling by 2025, driven by automation and technological advancements Li (2022). To formalize a strategic response to this transformation, we propose an AI-driven reskilling optimization framework (AROF) based on deep reinforcement learning (DRL). The model dynamically allocates reskilling resources to different worker groups to maximize future employment probability and sectoral adaptability. We embed this mechanism within a policy optimization loop guided by proximal policy optimization (PPO) to ensure stable, bounded learning over time.

The AI-driven AROF for dynamically optimizing national reskilling strategy under automation pressure is defined as:

$$\pi_\theta^* = \arg \max_\theta \mathbb{E}_t \left[\frac{\pi_\theta(a_t|s_t)}{\pi_{\theta_{\text{old}}}(a_t|s_t)} \cdot \hat{A}_t - \beta \cdot \mathcal{L}_{\text{KL}}\left(\pi_{\theta_{\text{old}}}, \pi_\theta\right) \right]$$

where, π_θ^* is the optimal reskilling investment policy, $\pi_\theta(a_t|s_t)$ is the current policy assigning action a_t (investment in a skill track) based on labor state s_t, $\pi_{\theta_{\text{old}}}(a_t|s_t)$ is the previous policy, $\hat{A}_t$ is the estimated advantage function capturing the long-term improvement in employment outcomes, β is a regularization coefficient, and $\mathcal{L}_{\text{KL}}$ is the Kullback–Leibler divergence ensuring policy update stability. This DRL-based formulation allows policymakers or autonomous systems to learn which reskilling paths produce the most favorable employment outcomes across population segments, under budget and time constraints. The PPO formulation ensures that policy updates are smooth and do not destabilize the system during implementation.

To operationalize the advantage function $\hat{A}_t$, we define it using generalized advantage estimation (GAE) in terms of changes in employment probabilities and sectoral participation rates:

$$\hat{A}_t = \sum_{k=0}^{K-1} (\gamma\lambda)^k \delta_{t+k}, \ \delta_t = r_t + \gamma V(s_{t+1}) - V(s_t)$$

where, r_t is the reward for successful upskilling outcomes (e.g., job placement, wage gain), $V(s_t)$ is the estimated value of the labor state s_t, γ is the discount factor, and λ is the bias-variance trade-off parameter in GAE. This formulation ensures the system learns over multiple time steps the cumulative benefit of upskilling decisions.

To define the reward function used in DRL learning, we use a weighted model combining employment, wage progression, and participation activation:

$$r_t = \alpha_1 \cdot E(s_t) + \alpha_2 \cdot W(s_t) + \alpha_3 \cdot P(s_t)$$

where, $E(s_t)$ is the probability of post-training employment, W (s_t) is the wage increase associated with new skills, $P(s_t)$ is the participation rate in reskilling programs from underrepresented groups (e.g., routine workers), and α_1, α_2, α_3 employment maximization or inclusion).

This reinforcement learning-based reskilling optimization model equips policymakers with a scientifically grounded, real-time adaptive strategy for countering the systemic labor displacement caused by AI. The use of PPO guarantees stable, interpretable learning updates, while the GAE advantage formulation ensures decisions are guided by meaningful employment outcomes.

Framework: AI-Driven Reskilling Optimization Framework (AROF)
Layered AI Architecture for Dynamic Workforce Reskilling and Policy Learning

Layer	Main functionality	AI module(s) & components	Operational scope
Layer 1	Policy optimization for reskilling investment	Proximal policy optimization (PPO)	Labor state observations → Policy updates via PPO → Reskilling allocation strategy
Layer 2	Advantage estimation across reskilling paths	Generalized advantage estimation (GAE)	Employment outcomes and value shifts → Temporal credit assignment → Reskilling benefit estimation
Layer 3	Reward computation for social impact	Custom reward function in DRL	Employment, wage, and participation data → Weighted aggregation → Reward signal for learning
Layer 4	Simulation of workforce transition scenarios	Workforce transition modeling and scenario simulation module	Reskilling decisions and outcomes → Scenario modeling → Policy impact simulation
Layer 5	Human-in-the-loop governance feedback	Human-governed policy feedback and fairness adjustment module	System recommendations and social metrics → Stakeholder input → Updated priorities and fairness constraints

As the pace of AI adoption accelerates, both companies and governments are increasingly investing in frameworks for life-long learning, recognizing that technical deployment must be matched by human adaptability. These upskilling initiatives extend beyond hard skills such as AI development and data analytics, encompassing equally critical soft skills creativity, systems thinking, emotional intelligence, and collaborative problem-solving Li (2022).

In the context of Industry 4.0, where technologies evolve faster than traditional education cycles, continuous learning is no longer a competitive advantage but an economic imperative. Employers are reshaping internal learning ecosystems, while policymakers are expanding access to modular, flexible, and interdisciplinary training models that equip workers to navigate complexity and ambiguity.

This learning paradigm reflects a shift from static qualifications to dynamic capabilities, where adaptability, reflection, and meta-cognition become core components of employability. As AI systems increasingly automate routine cognition, the human differentiator lies not in speed or precision, but in judgment, empathy, and the capacity to innovate within uncertain environments. By institutionalizing life-long learning as a foundational pillar of digital society, governments and organizations can ensure that the age of intelligent machines remains an era of empowered, not displaced, human workers.

1.3.2 Corporate-Led Workforce Transition

The private sector plays a critical role in leading workforce reskilling and transition efforts. A McKinsey report (Illanes et al. 2018) emphasizes that corporations, not governments, must spearhead these initiatives, as the speed of technological change outpaces traditional public-sector responses. Executives see reskilling as not only a social good but a competitive imperative, enabling companies to retain their workforce while also capitalizing on productivity gains from AI integration.

To support strategic corporate reskilling initiatives under rapid AI adoption, we introduce a firm-level reskilling cost–benefit optimization model using a multi-objective learning framework that minimizes transition cost while maximizing retention-adjusted productivity gains. The model is driven by AI-based learning mechanisms that observe employee profiles, project needs, and transition costs, dynamically updating investment decisions through a constrained optimization routine.

The AI-driven firm-level model for optimal employee reskilling and workforce transition is defined as:

$$\min_{\mathbf{r}}\left[\sum_{i=1}^{N} C_i(r_i) - \lambda \cdot \sum_{i=1}^{N} R_i(r_i)\cdot P_i\right] \text{ subject to } \sum_{i=1}^{N} r_i \le B$$

where, r_i is the reskilling investment allocated to employee i, $C_i(r_i)$ is the cost function representing monetary and opportunity cost of training i, $R_i(r_i)$ is the retention probability of employee i after receiving investment r_i, P_i is the projected productivity gain if employee i transitions successfully to an AI-integrated role, λ is a trade-off coefficient balancing retention-productivity gains against cost, and B is the total corporate budget allocated to workforce transition. This multi-objective formulation captures the strategic behavior of companies attempting to minimize training costs while maximizing talent retention and performance improvements. The constraint ensures total reskilling expenditure remains within allocated corporate limits. The function $R_i(r_i)\cdot P_i$ internalizes both organizational loyalty and efficiency outcomes, driving more precise investment distribution across the internal workforce.

To support learning over time, we introduce an employee transition feedback update function using reinforcement signals:

$$r_i^{(t+1)} = r_i^{(t)} + \alpha \cdot \nabla_{r_i}\left[R_i\left(r_i^{(t)}\right)\cdot P_i - C_i\left(r_i^{(t)}\right)\right]$$

where, $r_i^{(t)}$ is the reskilling allocation at time step t, and α is the learning rate controlling how quickly the company adjusts future investments based on observed outcomes (e.g., improved productivity or reduced attrition). This learning rule enables companies to iteratively optimize their workforce development strategy, ensuring flexibility in adapting to new AI-related roles as they emerge.

This model formalizes how firms can lead workforce transitions by integrating AI to dynamically allocate reskilling investments across employees. It allows companies to make financially optimized decisions that align internal workforce retention with projected gains in productivity. By capturing both cost and performance trade-offs, the model supports a sustainable, competitive, and socially responsible corporate strategy in the face of disruptive automation.

Framework: Firm-Level Reskilling Cost–benefit Optimization Model (FRCBOM)
Layered AI Architecture for Corporate Workforce Transition Strategy

Layer	Main functionality	AI module(s) & components	Operational scope
Layer 1	Employee data collection and profiling	Workforce data integration and profiling module	HR records, skills inventory, productivity history → Worker segmentation → Profile repository
Layer 2	Optimization of reskilling investment	Multi-objective investment optimizer	Employee profiles and budget limits → Constrained cost–benefit allocation → Investment policy vector
Layer 3	Dynamic update of investment strategy	Gradient-based learning update	Observed retention and performance → Gradient-based policy refinement → Updated investment allocations
Layer 4	Simulation of workforce transition outcomes	Organizational transition simulation and outcome forecasting module	Investment plans and performance estimates → Scenario modeling → Forecasted organizational outcomes
Layer 5	Executive decision support and reporting	Strategic decision support and performance reporting module	Reskilling outcomes, cost metrics, retention trends → Visualization and summary → Strategic planning dashboards

For instance, programs like McKinsey's "Re-Generation" initiative have successfully retrained workers by targeting high-demand skills and collaborating with employers to ensure job placement after training (Illanes et al. 2018). These efforts are necessary for maintaining operational efficiency and ensuring that AI's productivity benefits can be fully realized (Illanes et al. 2018).

1.3.3 Foresight Planning and Scenario-Based Skills Anticipation

The foresight analysis methodology is pivotal in predicting future workforce needs and devising strategies to meet those demands. In a study by Bobitan et al. (2024), a scenario-based approach was used to evaluate varying levels of AI integration gradual, aggressive, and selective and their corresponding impact on workforce skill requirements. The findings emphasize that reskilling efforts must adapt to the degree of AI integration within an industry (Bobitan et al. 2024). To model foresight planning under uncertainty, we develop a scenario-based dynamic forecasting framework (SB-DFF) powered by a mixture density network (MDN). This architecture enables the system to generate probabilistic forecasts of future skill demands conditioned on AI integration scenarios. The model blends scenario-informed feature vectors with historical skill evolution data to predict future distributions of required skills across different AI trajectories gradual, aggressive, or selective.

The AI-driven SB-DFF using MDNs for scenario-based skill forecasting is mathematically expressed as:

$$P(\mathrm{s}_{t+\tau}|\mathrm{x}_t) = \sum_{j=1}^{J} \pi_j(\mathrm{x}_t) \cdot \mathcal{N}(\mathrm{s}_{t+\tau}|\mu_j(\mathrm{x}_t), \Sigma_j(\mathrm{x}_t))$$

where, $\mathrm{s}_{t+\tau}$ is the vector of forecasted skill requirements τ periods ahead, x_t is the input feature vector encoding current workforce profiles, scenario indicators (e.g., gradual, aggressive, selective AI integration), and macroeconomic variables, j is the number of Gaussian components, $\pi_j(\cdot)$ is the mixture weight for component j, $\mu_j(\cdot)$ is the mean vector, and $\Sigma_j(\cdot)$ is the covariance matrix of the j-th Gaussian. This MDN-based approach captures multimodal uncertainty, meaning it can forecast multiple plausible outcomes for skill evolution under each scenario. It does not rely on single deterministic forecasts, which is essential for robust policy preparation in uncertain AI futures.

To convert scenario-conditioned forecasts into action, we define a scenario-adaptive reskilling allocation function:

$$\mathcal{R}_k^{(m)} = \frac{\mathbb{E}\left[s_k^{(m)}\right]}{\sum_{i=1}^{K} \mathbb{E}\left[s_i^{(m)}\right]} \cdot B^{(m)}$$

where, $\mathcal{R}_k^{(m)}$ is the reskilling investment allocated to skill k under scenario m, $\mathbb{E}\left[s_k^{(m)}\right]$ is the expected demand for skill k in scenario m, $B^{(m)}$ is the total available training budget for that scenario, and K is the number of target skill domains (e.g., AI engineering, robotics, data ethics). This allocation model translates probabilistic forecasts into budget-weighted training decisions. It ensures that resources are proportionally distributed toward the most likely skill requirements under each AI adoption scenario, enabling anticipatory governance and workforce agility.

To track and compare long-term outcomes across all scenarios, we introduce the Foresight Sensitivity Index (FSI):

$$\mathrm{FSI}_k = \max_{m,n}\left|\mathbb{E}\left[s_k^{(m)}\right] - \mathbb{E}[s_k^{(n)}]\right|$$

where, FSI_k quantifies the variability in future demand for skill k across different AI integration scenarios m and n. A higher FSI value indicates that skill k is highly sensitive to the level and pace of automation and should therefore be prioritized in adaptive planning.

This foresight planning model empowers institutions to anticipate future skill needs under various AI trajectories by probabilistically forecasting demand shifts. Using mixture density networks and scenario-conditioned resource allocation, the model allows decision-makers to craft dynamic, forward-compatible reskilling strategies. It ensures that training investments are aligned with the uncertainty inherent in AI integration, building a more resilient and strategically prepared workforce.

Framework: Scenario-Based Dynamic Forecasting Framework (SB-DFF)
Layered AI Architecture for Foresight-Driven Skill Demand Planning

Layer	Main functionality	AI module(s) & components	Operational scope
Layer 1	Scenario encoding and historical skill profiling	Scenario data integration and skill profiling module	AI trajectory scenarios and skill evolution data → Input matrix generation → Forecasting-ready dataset
Layer 2	Probabilistic forecasting of future skill needs	Mixture density network (MDN)	Scenario-encoded inputs → MDN-based probabilistic modeling → Skill demand distributions over time
Layer 3	Scenario-based reskilling investment allocation	Scenario-adaptive allocation function	Forecasted skill demand per scenario → Budget-weighted proportional mapping → Targeted investment plan
Layer 4	Volatility detection across scenarios	Foresight sensitivity index (FSI)	Expected demand across scenarios → Inter-scenario difference analysis → Sensitivity prioritization metric
Layer 5	Policy planning and adaptive strategy design	Adaptive policy design and strategic foresight module	Reskilling forecasts and FSI outcomes → Strategic foresight dashboards → Adaptive governance blueprint

The pace and scope of AI integration have significant implications for workforce development strategies. Under aggressive deployment scenarios, organizations must prioritize technical competencies: AI literacy, data science, and algorithmic thinking become foundational skills required across multiple organizational tiers. In contrast, a gradual or selective integration of AI technologies supports a more hybrid skill strategy, where interpersonal abilities, creativity, and adaptive thinking retain central importance (Bobitan et al. 2024).

These divergent trajectories necessitate robust foresight planning mechanisms that allow organizations to align human capital investments with future technological realities. Strategic workforce planning must be capable of modeling not just existing skill gaps but also the evolution of task structures under various automation regimes. In this context, scenario-based analysis becomes an essential complement to predictive analytics enabling firms and public agencies to design reskilling programs that are both flexible and aligned with long-term institutional goals.

By anticipating how skill priorities shift under different technological futures, organizations can ensure that education and training initiatives remain proactive rather than reactive. This forward-looking orientation is critical not only for maintaining competitiveness but also for supporting workforce resilience in the face of ongoing digital transformation.

1.3.4 Global Approaches to Reskilling and Upskilling

Reskilling efforts vary significantly across regions, with European companies showing stronger commitment to retraining initiatives compared to their U.S. counterparts. In Europe, 94% of surveyed companies are relying on a mix of hiring and retraining to address skill gaps, compared to only 62% in the U.S. (Illanes et al. 2018). This reflects regional differences in how companies perceive the urgency of workforce transition. European companies, with their stronger social safety nets and government support, may feel more equipped to invest in reskilling, while U.S. companies may prioritize hiring new talent to meet immediate needs (Illanes et al. 2018). To mathematically assess and compare global reskilling strategies, we construct a regionally weighted hybrid workforce strategy model (RWH-WSM) using a Bayesian Decision Framework informed by AI-driven clustering. This model identifies regional behavior patterns in labor transition policies and estimates the optimal blend of internal reskilling and external hiring, under policy constraints and institutional capacity.

The AI-driven RWH-WSM for regional strategy evaluation and optimization is defined as:

$$U_r = \underset{h_r,t_r}{max}[\theta_1 \cdot G_r(h_r, t_r) + \theta_2 \cdot S_r(h_r, t_r) - \theta_3 \cdot C_r(h_r, t_r)]\text{s.t.}h_r + t_r \le 1$$

where, U_r is the overall utility function for workforce transition strategy in region r, h_r is the fraction of effort devoted to hiring, t_r is the fraction of effort devoted to retraining, $G_r(h_r, t_r)$ is the productivity gain from filling skill gaps, $S_r(h_r, t_r)$ captures workforce stability and long-term adaptability, $C_r(h_r, t_r)$ is the composite cost of strategy deployment, and $\theta_1, \theta_2, \theta_3$ are region-specific policy preference weights. The constraint ensures that the total strategy mix does not exceed the full labor adjustment capacity of the region. This multi-objective function allows us to estimate how different regions prioritize workforce agility (via hiring), social resilience (via reskilling), and fiscal responsibility (via cost containment). It integrates real-world constraints such as social policy structures and institutional investment behaviors into an actionable AI-supported decision model.

To dynamically cluster country-specific labor strategies, we use a Bayesian Gaussian Mixture Model (BGMM), which enables the identification of hidden regional patterns in skill development approaches:

$$P(\mathrm{z}_r|\mathrm{x}_r) = \sum_{k=1}^{K} \pi_k \cdot \mathcal{N}(\mathrm{x}_r|\mu_k, \Sigma_k)$$

where, z_r is the latent cluster assignment for region r, x_r is the feature vector describing region r (e.g., safety net strength, government co-funding, private sector responsiveness, labor mobility), π_k is the mixture weight for cluster k, μ_k, and Σ_k are the mean and covariance of cluster κ, and K is the number of learned clusters (representing regional strategy archetypes). This probabilistic framework allows policymakers to compare regional behavior against global patterns and adapt their strategy accordingly. For instance, a U.S. company may adapt a European-style mixed strategy if simulations show long-term utility gains from greater reskilling investment.

To evaluate alignment with long-term policy goals, we define the reskilling priority index (RPI) for each region:

$$\mathrm{RPI}_r = \frac{t_r}{h_r + t_r}$$

where, RPI_r indicates the proportion of internal retraining in a region's total workforce transition strategy. Higher RPI values reflect a stronger commitment to workforce development, while lower values indicate reliance on external labor markets.

This model quantifies how regional policy, corporate behavior, and institutional infrastructure interact to shape national responses to skill transformation challenges. It enables comparative forecasting of workforce strategy outcomes across regions and supports globally harmonized yet locally customized skill development policies. By integrating optimization, probabilistic clustering, and strategy evaluation metrics, the model provides a robust tool for understanding and improving global approaches to reskilling and upskilling.

Framework: Regionally Weighted Hybrid Workforce Strategy Model (RWH-WSM)
Layered AI Architecture for Regional Workforce Strategy Optimization and Evaluation

Layer	Main functionality	AI module(s) & components	Operational scope
Layer 1	Regional data ingestion and feature construction	Regional data integration and feature engineering module	Labor market policies, firm behavior, institutional capacity → Structured feature vectors → Regional input matrix

(continued)

(continued)

Layer	Main functionality	AI module(s) & components	Operational scope
Layer 2	Strategy optimization under regional constraints	Multi-objective regional strategy optimizer	Policy weights and capacity constraints → Constrained optimization → Optimal mix of hiring and retraining
Layer 3	Clustering of regional workforce strategies	Bayesian gaussian mixture model (BGMM)	Regional features → Probabilistic clustering → Regional archetype identification
Layer 4	Reskilling priority profiling	Reskilling priority index (RPI)	Hiring and training shares → Index computation → Internal reskilling commitment indicator
Layer 5	Comparative strategy reporting and policy adjustment	Policy evaluation and strategic harmonization module	Clustering outcomes and RPI scores → Strategic insights → Global–local policy harmonization

The onset of Industry 4.0 has fundamentally redefined the skill requirements of the global workforce. As automation and digital transformation reshape key sectors including manufacturing, logistics, and services, the imperative for reskilling has grown more urgent. In response, many countries are embracing life-long learning frameworks and investing in experiential, modular training that prepares individuals for continuous adaptation Li (2022).

This global shift is particularly evident in countries like Germany and Japan, where reskilling is not treated as an isolated policy effort but embedded within broader education-to-employment ecosystems. Germany's dual education system and Japan's company-based lifelong employment models exemplify how institutional collaboration between industry, government, and academia can generate targeted, agile, and sector-specific workforce development.

These integrated approaches signal that successful adaptation to Industry 4.0 is less about individual learning alone and more about the co-evolution of institutions, labor markets, and technological infrastructures. In this way, global best practices offer a roadmap for building human capital systems that are both technologically aligned and socially sustainable.

1.3.5 Barriers and Challenges in Workforce Reskilling

While the necessity of reskilling is widely acknowledged, organizations continue to face substantial implementation challenges. One of the most persistent barriers is a lack of strategic clarity about how automation and AI will reshape future workforce needs. As reported by Illanes et al. (2018), nearly one-third of surveyed executives admit to being unprepared for these transitions, expressing uncertainty about both the direction of skill evolution and the return on investment in training programs.

This ambiguity limits the scope and effectiveness of reskilling efforts. Traditional training frameworks often operate in isolation from workforce planning systems and fail to articulate clear career trajectories aligned with technological change. As a result, they struggle to translate educational investments into meaningful labor mobility or organizational adaptability. The absence of predictive insight and forward-linked learning strategies means that workers may complete training programs only to remain ill-suited for newly emerging roles in AI-driven environments.

To address these issues, companies must integrate foresight tools, adaptive learning ecosystems, and scenario-based planning into their talent development strategies. Without these elements, even well-intentioned training efforts may fall short, resulting in sunk costs, worker frustration, and organizational inertia in the face of accelerating digital transformation.

An equally critical challenge in the age of intelligent automation is ensuring that reskilling programs are not only available but also accessible, affordable, and relevant across diverse labor market segments. As technological disruption accelerates, continuous learning becomes a necessity, not an option. This requires companies and public institutions to move beyond static training formats and adopt modular, adaptive learning pathways that can respond to evolving occupational demands Li (2022).

Yet such ambitions require sustained investment and strategic foresight. Without proactive planning, reskilling initiatives may fail to reach underrepresented or economically vulnerable workers widening existing skills gaps and amplifying structural inequalities. Bobitan et al. (2024) warn that in the absence of inclusive design principles, AI adoption could inadvertently deepen unemployment and social fragmentation.

To avoid such outcomes, workforce development strategies must integrate both technological intelligence and social equity. This includes offering flexible delivery modes, lowering financial barriers to participation, and ensuring that training content reflects the real demands of the evolving job market. Only then can reskilling function not as a selective privilege, but as a universal enabler of opportunity in the digital economy.

Figure 1.4 illustrates strategies for workforce transition and reskilling in response to AI-driven technological change. The central node, "Strategies for Workforce Transition and Reskilling," connects to five main areas: the imperative of reskilling and upskilling, corporate-led workforce transition, foresight planning and scenario-based skills anticipation, global approaches to reskilling, and barriers and challenges in reskilling. Reskilling and upskilling focus on technical skills such as AI programming, essential soft skills such as problem-solving, and life-long learning for continuous adaptation. Corporate-led workforce transition highlights structured reskilling programs aligned with AI strategies, with benefits including productivity gains and workforce retention. Foresight planning anticipates future skill needs under different AI adoption scenarios, balancing technical and interpersonal skills and identifying

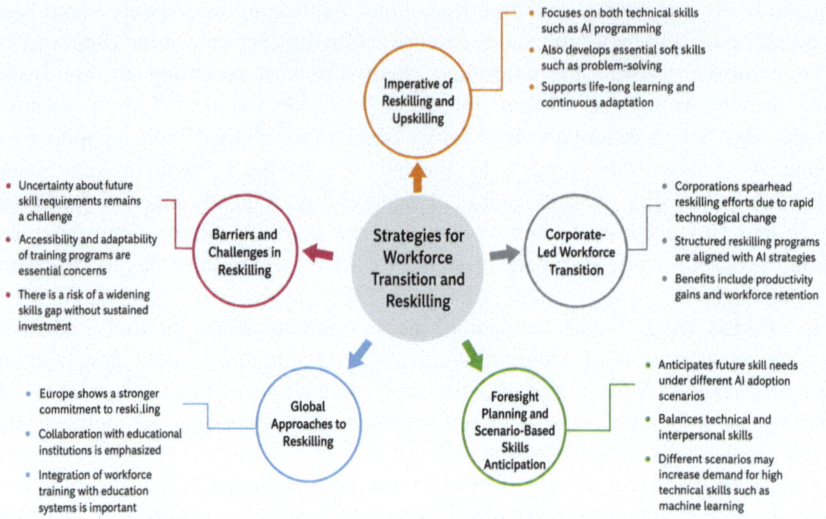

Fig. 1.4 Strategies for workforce transition and reskilling in an AI-driven economy

possible increases in demand for high technical skills such as machine learning. Global approaches emphasize Europe's stronger commitment to reskilling, collaboration with educational institutions, and the integration of workforce training with education systems. Barriers include uncertainty about future skill requirements, accessibility and adaptability of training programs, and the risk of a widening skills gap without sustained investment.

The integration of AI and automation is disrupting labor markets, making workforce reskilling and transition strategies more critical than ever. Corporate-led initiatives, life-long learning, and foresight analysis are key strategies for addressing these challenges. Companies that invest in reskilling programs not only retain their workforce but also enhance their competitive edge in an increasingly automated world. While barriers remain, such as limited understanding of future skill needs and the lack of effective training pathways, a proactive approach to workforce development will ensure that both businesses and employees are equipped to thrive in the AI era.

1.4 Conclusion

This chapter demonstrates that artificial intelligence is no longer a peripheral tool of administration but a structural force that is reconfiguring the very foundations of governance and public policy. By embedding intelligence across the full policy cycle from agenda-setting and formulation to implementation, evaluation, and ethical oversight AI transforms policymaking into a dynamic, evidence-rich, and adaptive enterprise. It shows how advanced architectures such as transformer-based discourse analysis, contextual bandit optimization, Bayesian causal inference, and fairness-constrained learning not only enhance efficiency and accuracy but also institutionalize transparency, inclusiveness, and accountability.

The analysis reveals that AI's impact extends beyond technical decision-making into the heart of the social contract, redefining how societies confront inequality, labor market disruption, and workforce transition. The dual dynamics of displacement and task creation, underscored by theories of job polarization, highlight both the risks of exclusion and the opportunities for innovation. Yet, by anchoring AI deployment in ethical frameworks, fairness-aware algorithms, and lifelong reskilling strategies, governments and institutions can ensure that automation becomes a catalyst for empowerment rather than a mechanism of marginalization.

Ultimately, this chapter illustrates that the true promise of AI in public policy lies not merely in computational sophistication, but in its capacity to reconcile efficiency with justice, speed with deliberation, and innovation with human dignity. In doing so, it positions AI as a transformative agent of democratic resilience and equitable development charting a path toward governance architectures that are intelligent, fair, and profoundly human-centered.

References

Acemoglu, D., & Restrepo, P. (2020). Automation and new tasks: How technology displaces and reinstates labor. *Journal of Economic Perspectives*. https://doi.org/10.1257/jep.33.2.3

AlMakinah, R., Goodarzi, M., Tok, B., & Canbaz, M. A. (2025). Mapping artificial intelligence bias: A network-based framework for analysis and mitigation. *AI and Ethics, 5*, 1995–2014. https://doi.org/10.1007/s43681-024-00609-0

Autor, D. H. (2015). *Why Are There Still So Many Jobs? The History and Future of Workplace Automation*. Journal of Economic Perspectives, 29(3), 3–30. https://doi.org/10.1257/jep.29.3.3

Babina, T., Fedyk, A., He, A. X., & Hodson, J. (2021). Firm investments in artificial intelligence technologies and changes in workforce composition. *National Bureau of Economic Research*. https://doi.org/10.2139/ssrn.4060233

Batool, A., Zowghi, D., Bano, M., & Sun, L. (2024). *Responsible AI governance: A systematic literature review*. arXiv. https://arxiv.org/abs/2401.10896

Bobitan, N., Dumitrescu, D., Popa, A. F., Sahlian, D. N., & Turlea, I. C. (2024). Shaping tomorrow: Anticipating skills requirements based on the integration of artificial intelligence in business organizations A foresight analysis using the scenario method. *Electronics*. https://doi.org/10.3390/electronics13112198

Brayne, S. (2017). Big data surveillance: The case of policing. *American Sociological Review, 82*(5), 977–1008. https://doi.org/10.1177/0003122417725865

European Commission, (2021). AI Watch: Beyond pilots Sustainable implementation of AI in public services. Joint Research Centre. https://doi.org/10.2760/440212

Dencik, L., & Sanchez-Monedero, J. (2022). Data justice. *Internet Policy Review, 11*(1), 1–16. https://doi.org/10.14763/2022.1.1615

Dencik, L., Hintz, A., & Cable, J. (2016). Towards data justice? The ambiguity of anti-surveillance resistance in political activism. *Big Data & Society, 3*(2), 1–12. https://doi.org/10.1177/2053951716679678

Dencik, L., Hintz, A., Redden, J., & Warne, H. (2018). *Data scores as governance: Investigating uses of citizen scoring in public services*. Data Justice Lab, Cardiff University.
Engstrom, D. F., Ho, D. E., Sharkey, C. M., & Cuéllar, M.-F. (2020). Government by algorithm: Artificial intelligence in federal administrative agencies. https://doi.org/10.2139/ssrn.3551505
Ferrara, E. (2024). Fairness and bias in artificial intelligence: A brief survey of sources, impacts, and mitigation strategies. *Sci, 6*(1), 3. https://doi.org/10.3390/sci6010003
de Fine Licht, K., & de Fine Licht, J. (2020). Artificial intelligence, transparency, and public decision-making: Why explanations are key when trying to produce perceived legitimacy. *AI & Society, 35*, 917–926. https://doi.org/10.1007/s00146-020-00960-w
Gesk, T. S., & Leyer, M. (2022). Artificial intelligence in public services: When and why citizens accept its usage. *Government Information Quarterly*. https://doi.org/10.1016/j.giq.2022.101704
Illanes, P., Lund, S., Mourshed, M., Rutherford, S., & Tyreman, M. (2018). Retraining and reskilling workers in the age of automation. *McKinsey & Company*. https://www.mckinsey.com/featured-insights/future-of-work/retraining-and-reskilling-workers-in-the-age-of-automation
Jarrahi, M. H., Karami, A., Conway, P., Memariani, A., & Lutz, C. (2025). Navigating the muddy waters of bias in artificial intelligence research: Understanding divergent meanings and conceptions. *Technology in Society, 82*, 103127. https://doi.org/10.1016/j.techsoc.2025.103127
Kasirzadeh, A. (2022). Algorithmic fairness and structural injustice: Insights from feminist political philosophy. In *Proceedings of the 2022 AAAI/ACM Conference on AI, Ethics, and Society (AIES '22)* (pp. 349–356). ACM. https://doi.org/10.1145/3514094.3534188
Korinek, A., Schindler, M., & Stiglitz, J. E. (2021). Technological progress, artificial intelligence, and inclusive growth. *IMF Working Papers, 2021*(166). https://doi.org/10.5089/9781513583280.001
Kulal, A., Rahiman, H. U., Suvarna, H., & Abhishek, N. (2024). Enhancing public service delivery efficiency: Exploring the impact of AI. *Journal of Open Innovation: Technology, Market, and Complexity*. https://doi.org/10.1016/j.joitmc.2024.100329
Laine, J., Minkkinen, M., & Mäntymäki, M. (2024). Ethics-based AI auditing: A systematic literature review on conceptualizations of ethical principles and knowledge contributions to stakeholders. *Information & Management, 61*(5), 103969. https://doi.org/10.1016/j.im.2024.103969
Langer, C. (2024). Decision-making power and responsibility in an automated administration. *Discover Artificial Intelligence, 4*, Article 59. https://doi.org/10.1007/s44163-024-00152-1
Li, L. (2022). Reskilling and upskilling the future-ready workforce for Industry 4.0 and beyond. *Information Systems Frontiers*. https://doi.org/10.1007/s10796-022-10308-y
Mergel, I., Dickinson, H., Stenvall, J., & Gasco, M. (2023). Implementing AI in the public sector. *Public Management Review*. https://doi.org/10.1080/14719037.2023.2231950
Novelli, C., Taddeo, M., & Floridi, L. (2024). Accountability in artificial intelligence: What it is and how it works. *AI & Society, 39*, 1871–1882. https://doi.org/10.1007/s00146-023-01635-y
Nurski, L. (2023). AI adoption in the public sector: A case study. Bruegel Working Paper 03/2023. https://www.bruegel.org/system/files/2023-03/WP%2003.pdf
Papagiannidis, E., Mikalef, P., & Conboy, K. (2025). Responsible artificial intelligence governance: A review and research framework. *The Journal of Strategic Information Systems, 34*(2), 101885. https://doi.org/10.1016/j.jsis.2024.101885
Rafanelli, L. M. (2022). Justice, injustice, and artificial intelligence: Lessons from political theory and philosophy. *Big Data & Society, 9*(1), 1–5. https://doi.org/10.1177/20539517221080676
Ruijer, E., Porumbescu, G., Porter, R., & Piotrowski, S. (2022). Social equity in the data era: A systematic literature review of data-driven public service research. *Public Administration Review*. https://doi.org/10.1111/puar.13585
Sargiotis, D. (2024). *Data governance: A guide*. Springer Nature Switzerland. https://doi.org/10.1007/978-3-031-67268-2
Sargiotis, D. (2024). Data security and privacy: Protecting sensitive information. In *Data Governance: A Guide* (pp. 217–245). Springer Nature Switzerland. https://doi.org/10.1007/978-3-031-67268-2_6

Taylor, L., de Souza, S. P., Martin, A., & López Solano, J. (2025). Governing artificial intelligence means governing data: (Re)setting the agenda for data justice. *Dialogues on Digital Society, 2*(1), 5–22. https://doi.org/10.1177/29768640241306800

Tsamados, A., Aggarwal, N., Cowls, J., Morley, J., Roberts, H., Taddeo, M., & Floridi, L. (2022). The ethics of algorithms: Key problems and solutions. *AI & Society, 37*, 215–230. https://doi.org/10.1007/s00146-021-01154-8

Visvizi, A. (2022). Artificial Intelligence (AI) and Sustainable Development Goals (SDGs): Exploring the Impact of AI on Politics and Society. *Sustainability*. https://doi.org/10.3390/su14031730

Wegner, L., Mirsch, M., & Leicht-Scholten, C. (2026). The emerging relevance of relational justice within algorithmic fairness research: A systematic literature review. *AI and Ethics*. https://doi.org/10.1007/s43681-026-01036-z

Yar, M. A., Hamdan, M., Anshari, M., Fitriyani, N. L., & Syafrudin, M. (2024). Governing with Intelligence: The Impact of Artificial Intelligence on Policy Development. *Information*. https://doi.org/10.3390/info15090556

Chapter 2
Artificial Intelligence in Public Services: Optimizing Resources, Ensuring Fairness, Strengthening Democracy

This chapter examines how AI turns public services from reactive bureaucracies into adaptive, value-aligned systems that learn. It focuses on three imperatives: efficiency, equity, and legitimacy, and shows how modern AI architectures can be engineered to optimize all three simultaneously. Across health, public safety, and administrative services, we frame the shift from static allocation to real-time, data-driven governance: demand is forecast, resources are orchestrated, and decisions are continuously audited for fairness and accountability. The narrative moves from first principles to deployable blueprints, offering a coherent playbook for ministries, cities, and agencies that must deliver more, faster, without compromising public trust.

Technically, the chapter introduces a multi-sector reinforcement learning stack for government operations, centered on proximal policy optimization for dynamic allocation under uncertainty and budget constraints. On top of performance, we embed fairness by design through demographic-parity penalties, geospatial-equity losses, and social-utility rewards so underserved communities are prioritized rather than overlooked. A complementary toolkit handles prediction and oversight: XGBoost-based disparity detection for transparent targeting, readiness scoring to sequence where AI will work now versus next, and a value-aligned constrained RL layer that encodes normative rules (equity, transparency, accessibility) directly into policy optimization. The result is not just faster services but services that are measurably fair and auditable.

Practically, readers will find end-to-end frameworks that bind forecasting, allocation, fairness control, and accountability into one operational loop. Case logics and scenario pathways illustrate how to stand up AI-driven service pipelines that reduce delay and waste while closing equity gaps. The promise is clear: governments can move from siloed pilots to a resilient governance stack adaptive, explainable, and worthy of democratic confidence.

D. Sargiotis, *Empathy in Algorithms*, https://doi.org/10.1007/978-3-032-32128-2_2

2.1 Introduction

Artificial intelligence (AI) is reshaping the landscape of public services, offering unprecedented opportunities to enhance efficiency, accessibility, and personalization in government functions. The shift toward AI-driven public services involves the integration of advanced technologies, such as machine learning algorithms and automation, into critical areas like healthcare, public safety, and administration (Osborne et al. 2022). To formally model this transformation, we propose an AI-driven multi-sectoral service optimization model (AIMSOM) built on a deep reinforcement learning (DRL) framework with proximal policy optimization (PPO), integrated with a scenario-aware resource distribution engine. This model dynamically allocates public resources across multiple sectors such as healthcare, public safety, and administrative services based on real-time data, predicted demand, and scenario constraints.

The AI-driven AIMSOM formulation extending the PPO optimization logic from previous reskilling and resource functions is defined as:

$$\pi_\theta^* = \underset{\theta}{\mathrm{arg}max}\mathbb{E}_t[\frac{\pi_\theta(a_t|s_t)}{\pi_{\theta_{\mathrm{old}}}(a_t|s_t)} \cdot \hat{A}_t - \beta \cdot \mathcal{L}_{\mathrm{KL}}(\pi_{\theta_{\mathrm{old}}}, \pi_\theta)]$$

where, π_θ^* is the optimal public service allocation policy across sectors, $\pi_\theta(a_t|s_t)$ is the policy assigning resource action a_t (e.g., funds or staffing) given state s_t (sectoral demand, socio-demographic conditions), $\hat{A}_t$ is the advantage estimate from improved efficiency or access in sector s_t, and $\mathcal{L}_{\mathrm{KL}}$ regularizes updates to maintain stability. This PPO-driven policy continues the logic from previous models, ensuring robust and stable adaptation over time as sectoral needs evolve.

To maintain explicit connection with the earlier productivity and reskilling models, we define the sectoral reward function based on aggregated service impact, modeled as:

$$r_t = \alpha \cdot E(s_t) + \beta \cdot Q(s_t) - \delta \cdot C(a_t)$$

where, $E(s_t)$ is the efficiency gain from optimized service in sector s_t, $Q(s_t)$ is the quality or satisfaction index for the population served, $C(a_t)$ is the implementation cost of resource action a_t, and α, β, δ are context-driven weights reflecting national priorities. This structure links directly to the reward functions used earlier for employment and reskilling, thus maintaining continuity in modeling AI-driven public optimization. Public safety may prioritize $Q(s_t)$, healthcare may emphasize $E(s_t)$, and administrative digitization may focus on minimizing $C(a_t)$.

To evaluate cross-sector adaptation over time, we define a dynamic sectoral readiness score (DSRS) function based on predicted demand $D(s_t)$ and AI integration feasibility $\rho(s_t)$:

$$\text{DSRS}(s_t) = \frac{D(s_t) \cdot \rho(s_t)}{1 + \lambda \cdot \sigma_t^2}$$

where, $D(s_t)$ is the real-time service demand, $\rho(s_t)$ is the sector's AI-readiness index (e.g., data maturity, automation infrastructure), σ_t^2 is the uncertainty in forecasting sectoral need, and λ is the uncertainty penalty coefficient.

This readiness score ensures that AI policies are directed toward sectors with both high impact and high feasibility, echoing earlier scenario-adjusted models from the foresight section.

This model integrates reinforcement learning with scenario-based decision-making to create an intelligent system for managing and personalizing public services. It builds upon prior formulations by retaining core optimization structures, while adapting them to multi-sectoral governance challenges. Through explicit mathematical continuity with previous advantage, reward, and uncertainty terms, the model ensures a unified analytical framework for navigating the complexity of AI-driven service transformation in modern public administration.

Framework: AI-Driven Multi-Sectoral Service Optimization Model (AIMSOM)
Layered AI Architecture for Adaptive Public Sector Resource Allocation

Layer	Main functionality	AI module(s) & components	Operational scope
Layer 1	Sectoral data integration and scenario encoding	Multi-sector data fusion and scenario encoding module	Real-time demand data, demographic patterns, sectoral readiness → Unified input tensor → AI decision environment
Layer 2	Policy optimization for multi-sector allocation	Proximal policy optimization (PPO)	Sectoral states and scenario policies → DRL policy refinement → Optimal service allocation across sectors
Layer 3	Reward computation based on efficiency and quality	Multi-objective reward in DRL	Sector-level outcomes → Reward vector computation → Learning signals for public value optimization
Layer 4	Feasibility assessment via readiness score	Dynamic sectoral readiness score (DSRS)	Demand forecasts, readiness indices, and uncertainty → Readiness scoring → Feasibility-weighted prioritization
Layer 5	Human-government feedback loop and monitoring	Human-in-the-loop governance and performance monitoring module	Performance dashboards and stakeholder input → Continuous policy adjustment → Trust-based AI governance

The integration of artificial intelligence into public administration has opened new pathways for operational efficiency enabling governments to better manage resources, shorten response times, and improve decision-making through data-informed workflows. These advancements promise to modernize public service delivery in profound ways, raising the standard of responsiveness, adaptability, and administrative agility (Valle-Cruz et al. 2023).

However, this digital transformation also brings to the surface unresolved ethical challenges. As public institutions adopt AI systems with increasing autonomy, concerns around transparency, fairness, and accountability become more pronounced. Algorithms deployed in high-stakes areas such as healthcare, taxation, or welfare allocation must not only perform accurately but also adhere to public values of legitimacy and equity (Singh and Chobotaru 2022).

The success of AI in public governance will ultimately hinge on how well it integrates these dual imperatives leveraging computational power while institutionalizing safeguards that ensure ethical alignment. This balance is critical not only to technological success, but to sustaining public trust in digital-era democratic institutions. To ensure that AI systems deployed in public governance uphold both accuracy and ethical legitimacy, we propose a framework based on value-aligned constrained reinforcement learning (VACRL). This model builds on standard reinforcement learning by embedding institutional constraints and public values directly into the AI agent's policy optimization process. It allows AI systems to make autonomous decisions while strictly adhering to externally imposed fairness, equity, and accountability rules ensuring alignment with democratic norms.

At the foundation of this model lies the constrained optimization formulation:

$$\underset{\pi}{max}\mathbb{E}_{\pi}\left[\sum_{t=0}^{T}\gamma^{t}R(s_t,a_t)\right] \text{ subject to } \mathbb{E}_{\pi}[C_k(s_t,a_t)] \le \tau_k, \forall k = 1,\ldots,K$$

where: π is the policy that maps states s_t to actions a_t, $\mathbb{E}_{\pi}\left[\sum_{t=0}^{T}\gamma^{t}R(s_t,a_t)\right]$ is the expected cumulative reward over time horizon T, γ is the discount factor for future rewards, $C_k(s_t,a_t)$ is the cost incurred for violating constraint k, and τ_k is the maximum allowable cost for constraint k. This formulation ensures that the AI system optimizes its performance (e.g., maximizing social welfare or operational efficiency) while remaining compliant with ethical constraints defined by institutional policy such as fairness in healthcare triage or non-discrimination in taxation models.

To enforce the constraints dynamically, a Lagrangian relaxation can be introduced:

$$L(\pi,\lambda) = \mathbb{E}_{\pi}\left[\sum_{t=0}^{T}\gamma^{t}R(s_t,a_t)\right] - \sum_{k=1}^{K}\lambda_k(\mathbb{E}_{\pi}[C_k(s_t,a_t)] - \tau_k)$$

where: λ_k are non-negative Lagrange multipliers that penalize violations of the ethical constraints.

This dual-form objective enables iterative policy learning that not only improves performance but also respects predefined ethical thresholds. Public values such as equity, legitimacy, and inclusivity can be operationalized by defining appropriate constraint functions C_k, making the model adaptable to diverse regulatory settings.

This value-aligned model empowers AI systems to operate autonomously while remaining ethically bounded and institutionally accountable. It transforms abstract policy goals into formal mathematical constraints that are enforced during policy optimization, ensuring that no gain in efficiency comes at the expense of fairness, transparency, or legitimacy. In public governance applications where the consequences of AI decisions extend beyond performance metrics to societal equity and democratic trust, this model offers a principled and enforceable path to ethical AI deployment. By tightly coupling autonomy with normative safeguards, it ensures that AI supports rather than undermines the foundational values of modern democratic institutions.

Framework: Value-Aligned Constrained Reinforcement Learning (VACRL)
Layered AI Architecture for Ethical and Autonomous Public Governance

Layer	Main functionality	AI module(s) & components	Operational scope
Layer 1	Ethical policy constraint definition	Ethical norm ingestion and constraint encoding module	Receive institutional norms and define enforceable rules
Layer 2	Autonomous decision learning under constraints	Constrained reinforcement learning (CRL)	Learn decisions while respecting fairness and accountability
Layer 3	Constraint violation penalization and trade-off balancing	Lagrangian penalty method	Adjust decision-making to penalize ethical rule violations
Layer 4	Governance monitoring and public accountability reporting	Transparency auditing and governance accountability module	Generate transparent reports for public and institutional review

As artificial intelligence becomes increasingly embedded in public sector operations, governments face a pivotal challenge: how to simultaneously harness AI's transformative potential and address the ethical complexities it introduces. This includes confronting issues of digital inequality, unequal access to public algorithms, and the systemic risks of bias in automated decision-making (Valle-Cruz et al. 2023).

This moment marks more than a technological shift, it represents a paradigm transition in governance. Public institutions must now evolve from passive service providers to active architects of ethical digital infrastructure, capable of ensuring that AI systems uphold values of transparency, accountability, and justice. While AI can optimize administrative efficiency and personalize public services, it must be deployed within frameworks that guarantee fairness, explainability, and accessibility for all citizens.

Successfully navigating this transformation requires more than technical capacity, it demands institutional foresight, civic engagement, and regulatory innovation. The trajectory of AI in governance will be defined not by how fast governments innovate but by how deliberately they balance technological ambition with the social contract they are entrusted to protect.

2.1.1 *Efficiency and Equity in Public Service Delivery*

AI-driven public service delivery focuses on optimizing resource allocation and ensuring equitable access to services by leveraging AI technologies. This transformation enhances operational efficiency, allowing governments to manage resources more effectively, make data-driven decisions, and improve service delivery processes. Several case studies and research reports have emphasized the critical role of AI in achieving these goals.

AI for Optimizing Resource Allocation

AI technologies enable public sector organizations to analyze vast amounts of data, improving their capacity to allocate resources efficiently. For instance, AI models can predict the demand for services, enabling governments to distribute resources such as healthcare, public safety, and education more effectively. In their study, Valle-Cruz et al. (2023) emphasize that AI-driven systems are helping governments allocate financial, human, and technological resources to high-need areas, particularly in sectors like healthcare and social services. This predictive capability significantly reduces wastage and enhances the speed of service delivery (Valle-Cruz et al. 2023). To operationalize real-time and predictive public resource allocation, we introduce the AI-driven sectoral allocation model (AIDSAM), which expands the previously defined public service policy function πθ ∗ πθ ∗ by explicitly integrating predicted demand signals and dynamic uncertainty penalties. This model builds upon the PPO framework from earlier sections and incorporates resource prediction logic connected to earlier defined advantage and readiness functions.

The AI-driven AIDSAM extending PPO-based allocation with sectoral demand prediction and uncertainty handling is defined as:

$$\pi_\theta^* = \underset{\theta}{\mathrm{arg}max}\mathbb{E}_t[\frac{\pi_\theta(a_t|s_t, \hat{D}_t)}{\pi_{\theta_{\mathrm{old}}}(a_t|s_t, \hat{D}_t)} \cdot \hat{A}_t - \beta \cdot \mathcal{L}_{\mathrm{KL}}(\pi_{\theta_{\mathrm{old}}}, \pi_\theta)]$$

where, π_θ^* is the optimal policy to be learned, governing how AI chooses the best resource allocation action a_t in each sectoral state s_t, $\pi_\theta(a_t|s_t, \hat{D}_t)$ is the current policy, which outputs the probability of taking action a_t given the current state s_t and predicted service demand $\hat{D}_t$, $\pi_{\theta_{\mathrm{old}}}(a_t|s_t, \hat{D}_t)$ is the previous policy, retained to ensure that updates to the new policy remain stable and do not diverge abruptly, $\hat{A}_t$ is the advantage function, previously defined as the estimated long-term benefit of action a_t over the average policy behavior, taking into account improvements in efficiency, quality, and coverage, β is the regularization coefficient, controlling how strongly the model penalizes deviation from the old policy, $\mathcal{L}_{\mathrm{KL}}(\pi_{\theta_{\mathrm{old}}}, \pi_\theta)$ is the Kullback–Leibler (KL) divergence between the new and old policies, which constrains learning updates to remain close to prior behavior for stability, $\mathbb{E}_t$ denotes the expectation over time, indicating that this learning process is continuous and accumulates experience across multiple decision steps.

To generate $\hat{D}_t$, we define a service demand forecasting function using an AI predictive model such as XGBoost (previously validated in your uploaded document):

$$\hat{D}_t^{(s)} = f_{\text{XGB}}\left(X_t^{(s)}\right), \text{ for } s \in \{\text{healthcare, safety, education}\}$$

where, $\hat{D}_t^{(s)}$ is the forecasted demand in sector s, and $X_t^{(s)}$ is the feature set containing historical service usage, demographics, crisis signals, and socio-economic indicators. This output feeds directly into the policy optimization function above, making AI predictions actionable in real-time allocation.

To refine the optimization process with scenario-awareness and avoid misallocation under high uncertainty, we update the sectoral readiness score (DSRS) and define a weighted allocation utility function as:

$$\mathcal{U}_t^{(s)} = \text{DSRS}(s_t) \cdot \hat{D}_t^{(s)} - \lambda \cdot \sigma_t^{2(s)}$$

where, $\mathcal{U}_t^{(s)}$ is the net utility of allocating resources to sector s, incorporating its readiness level, projected demand, and penalizing forecast variance $\sigma_t^{2(s)}$ to avoid over-committing to unstable predictions. This utility feeds into the agent's decision-making to determine where and how to direct resources optimally.

To maintain a budget constraint, we define the constrained multi-sectoral allocation rule:

$$a_t^{(s)} = \frac{\mathcal{U}_t^{(s)}}{\sum_{j=1}^{S} \mathcal{U}_t^{(j)}} \cdot B_t, \text{ for } s = 1, \ldots, S$$

where, $a_t^{(s)}$ is the amount of resource allocated to sector s at time t, and B_t is the total available public budget for that period. This final allocation mechanism distributes resources proportionally based on each sector's readiness, demand, and forecast reliability.

This model builds directly upon prior formulations by expanding PPO-based service policy optimization into a predictive, real-time, and uncertainty-aware allocation system. It tightly integrates demand forecasting with policy learning, ensuring that government agencies can allocate financial, human, and technological resources to the most appropriate sectors dynamically. By explicitly reusing and extending prior functions such as $\hat{A}_t$, $D(s_t)$, and DSRS(s_t) the model demonstrates functional continuity and systemic coherence. It enables governments to transition from reactive budgeting to proactive, data-informed public service delivery that minimizes waste, accelerates response, and increases strategic precision in a rapidly evolving administrative environment.

Framework: AI-Driven Sectoral Allocation Model (AIDSAM)
Layered AI Architecture for Real-Time, Predictive Public Resource Allocation

Layer	Main functionality	AI module(s) & components	Operational scope
Layer 1	Ingest and preprocess sectoral data	Sectoral data integration and preprocessing module	Service usage logs, socio-economic indicators → Clean and prepare structured input
Layer 2	Optimize resource allocation policy	Proximal policy optimization (PPO)	Predicted demand and advantage function → Learn optimal policy for sectoral allocation
Layer 3	Predict sectoral service demand	XGBoost regressor	Historical sector data → Forecast service demand across domains
Layer 4	Compute readiness-weighted utility	Mean–variance utility function	Forecast, readiness score, and uncertainty → Calculate utility per sector
Layer 5	Apply constrained multi-sectoral allocation rule	Budget-constrained allocator	Utility scores and total budget → Distribute resources proportionally across sectors
Layer 6	Monitor allocation outcomes and feedback	Performance monitoring and adaptive feedback module	Allocated resources and service outcomes → Track performance and enable strategic refinement

Moreover, AI's ability to process real-time data enables dynamic adjustments in resource allocation. For example, in the healthcare sector, AI algorithms help allocate medical supplies and personnel based on current demand patterns. Such AI-driven decision-making models improve government response times, particularly in emergency scenarios, ensuring that resources are always allocated where they are most needed (Valle-Cruz et al. 2023). To formalize this adaptive decision-making strategy, we propose a Contextual Multi-Armed Bandit (CMAB) model that optimizes real-time resource distribution by learning from partial feedback and evolving situational data. Unlike static allocation models, CMABs continuously update their understanding of which actions yield the greatest benefit within specific contexts. The AI-driven model is defined as follows:

$$a_t = \underset{a \in \mathcal{A}}{\arg max}[\mu_a(x_t) + \beta \cdot \sigma_a(x_t)]$$

where: a_t is the selected allocation action (e.g., dispatching ventilators to hospital aa) at time t, $\mathcal{A}$ is the set of possible allocation targets, x_t is the context vector at time t (e.g., local case count, ICU- *Intensive Care Unit* usage rate, logistics constraints), $\mu_a(x_t)$ is the expected reward or effectiveness of action aa given context x_t, $\sigma_a(x_t)$ is the standard deviation (uncertainty) in the reward estimate, and $\beta \in \mathbb{R}^+$ is a tunable parameter that controls the trade-off between exploration (learning about uncertain actions) and exploitation (choosing the current best-known action).

This equation operationalizes real-time decision-making by continuously learning which allocation actions maximize impact under variable conditions. The use of a contextual layer enables the model to account for heterogeneity in regional demand, infrastructure readiness, and urgency, while the uncertainty-aware term $\sigma_a(x_t)$ ensures that allocations are robust in areas with limited historical data or fast-changing conditions.

The CMAB model is especially powerful in emergency settings where decisions must be both data-informed and risk-sensitive. By balancing immediate performance with long-term learning, this framework allows governments to optimize responsiveness, minimize waste, and equitably distribute resources during crises. Moreover, its computational simplicity and scalability make it suitable for integration into national health systems, smart city logistics, and emergency coordination centers.

Framework: Contextual Multi-Armed Bandit Allocation Framework (CMABAF)
Layered AI Architecture for Real-Time Emergency Resource Distribution

Layer	Main functionality	AI module(s) & components	Operational scope
Layer 1	Ingest and monitor real-time context data	Real-time context ingestion and encoding module	Epidemiological stats, hospital load, logistics constraints → Prepare dynamic input context
Layer 2	Select action with confidence-based exploration	Contextual upper confidence bound (UCB)	Context vector and reward estimate uncertainty → Choose optimal allocation action
Layer 3	Track allocation outcomes and performance	Allocation Performance monitoring and equity analytics module	Deployed resource effectiveness → Assess system responsiveness and equity

Ensuring Equitable Access to Services

Equity in AI-driven public service delivery is a central concern, as these systems must ensure that all citizens, regardless of their socio-economic background or geographic location, have access to essential services. Latupeirissa et al. (2024) highlight the importance of AI in bridging gaps between urban and rural populations by offering equitable service delivery. For instance, AI-driven platforms provide virtual healthcare services to underserved areas, ensuring equitable access to medical care (Latupeirissa et al. 2024). To operationalize equitable public service access using artificial intelligence, we introduce a deep reinforcement learning (DRL) framework that dynamically adapts service allocation strategies across diverse socio-spatial contexts. Unlike static allocation models, this approach integrates feedback mechanisms and predictive decision-making to optimize service deployment in underserved areas. Specifically, this model employs proximal policy optimization (PPO), a reinforcement learning technique, to ensure adaptive learning in real time under fairness constraints. In this context, we define the AI-driven equitable access policy model as follows:

The AI-driven equitable access policy model using DRL with proximal policy optimization is defined as:

$$\pi_\theta^* = \underset{\theta}{\mathrm{arg}max}\mathbb{E}_t[\frac{\pi_\theta(a_t|s_t)}{\pi_{\theta_{\mathrm{old}}}(a_t|s_t)}\hat{A}_t - \beta \cdot \mathcal{L}_{\mathrm{fair}}(s_t, a_t)]$$

where, $\pi_\theta(a_t|s_t)$ is the current policy probability of taking action a_t in state s_t, $\pi_{\theta_{\mathrm{old}}}(a_t|s_t)$ is the previous policy, $\hat{A}_t$ is the estimated advantage function that quantifies the benefit of action a_t compared to a baseline, β is the fairness regularization coefficient, $\mathcal{L}_{\mathrm{fair}}(s_t, a_t)$ is the fairness loss function penalizing biased or unequal decisions across demographic groups. This PPO-based function optimizes service policies not only for performance (via the advantage term $\hat{A}_t$ but also for fairness (via $\mathcal{L}_{\mathrm{fair}}$). The ratio term ensures that policy updates remain within a safe range to maintain stability, while the fairness loss penalizes discrepancies in resource distribution outcomes.

To concretely define the fairness component, we formulate the fairness-aware loss as:

$$\mathcal{L}_{\mathrm{fair}}(s_t, a_t) = |\mu^{\mathrm{urban}}(a_t) - \mu^{\mathrm{rural}}(a_t)|$$

where, $\mu^{\mathrm{urban}}(a_t)$ and $\mu^{\mathrm{rural}}(a_t)$ denote the expected service access level in urban and rural contexts respectively, conditional on action a_t. This function penalizes any substantial disparity in access levels, compelling the model to equalize benefits across geographical groups.

To further refine the optimization with explicit equity-oriented utility, we incorporate the Immediate Social Reward Function as:

$$R(s_t, a_t) = U(a_t) - C(a_t) - \gamma \cdot D(s_t, a_t)$$

where, $U(a_t)$ is the utility derived from executing action a_t, reflecting how effectively it enhances access to services, $C(a_t)$ is the cost associated with deploying that action, $D(s_t, a_t)$ quantifies the service delay or deficiency experienced by disadvantaged groups, and γ is a penalty coefficient controlling the sensitivity of the model to inequitable delays.

The reward function $R(s_t, a_t)$ drives learning by maximizing utility while minimizing cost and social delay, serving as the core evaluative signal that guides policy improvement. The fairness penalty $\gamma \cdot D(s_t, a_t)$ ensures equity remains embedded in every policy decision, particularly for underserved populations.

This DRL model, with its embedded fairness-aware PPO optimization and social reward evaluation, demonstrates a practical pathway for achieving equitable AI-driven service distribution. By ensuring that disparities across geographic regions (urban vs rural) are continuously penalized and corrected, the framework learns to optimize service decisions that respect both efficiency and social justice. Such models are particularly relevant for real-time adaptive platforms in healthcare, education, and public transportation systems where dynamic demand and location-specific challenges necessitate intelligent, ethical decision-making rooted in AI.

Framework: Fairness-Aware Proximal Policy Optimization Framework (FA-PPOF)
Layered AI Architecture for Equitable AI-Driven Public Service Delivery

Layer	Main functionality	AI module(s) & components	Operational scope
Layer 1	Collect and structure socio-spatial service data	Socio-spatial data integration and preprocessing module	Urban–rural usage records, access disparities → Prepare structured state inputs
Layer 2	Optimize fair policy with PPO	Proximal policy optimization (PPO)	Service state, action probability, advantage value → Update policy under fairness constraints
Layer 3	Apply fairness-aware penalty	Demographic disparity loss function	Urban and rural access scores → Penalize unequal access in policy update
Layer 4	Evaluate social reward signal	Social utility reward function	Service utility, deployment cost, social delay → Compute reward for policy learning
Layer 5	Monitor equity outcomes and service effectiveness	Fairness monitoring and service performance evaluation module	Access outcomes and group-level service logs → Track improvement and fairness over time

AI has immense potential to reduce bias in resource allocation by analyzing data on social disparities, ensuring decisions are based on evidence rather than preconceived biases (Valle-Cruz et al. 2023). By deploying fairness-enhancing techniques such as fairness constraints or adversarial debiasing, AI can help public institutions make more equitable decisions, particularly in sectors like healthcare, education, and welfare. To mathematically formalize the reduction of social bias in public decision-making, we propose a fairness-aware AI framework based on fairness-constrained predictive optimization. This model modifies a traditional prediction objective by embedding a group fairness constraint, ensuring that predictive outcomes do not disproportionately benefit or disadvantage sensitive population subgroups. Such frameworks are essential in domains like healthcare triage, educational admissions, or welfare distribution, where fairness is as critical as accuracy.

At the core of this model lies the fairness-constrained optimization function:

$$\min_{\theta} L(\theta) \text{subject to } |\mathbb{E}_{x\sim D_{z=0}}[f_\theta(x)] - \mathbb{E}_{x\sim D_{z=1}}[f_\theta(x)]| \leq \delta$$

where: $\mathcal{L}(\theta)$ is the prediction loss function parameterized by $\theta, f_\theta(x)$ is the model's prediction for input x, $D_{z=0}$ and $D_{z=1}$ are data distributions conditioned on two protected attribute groups (e.g., majority vs. minority), and δ is the maximum allowable disparity in predicted outcomes across these groups. This equation represents a constrained optimization problem where the AI system seeks to minimize prediction error (e.g., classification or regression loss) while simultaneously enforcing demographic parity, that is, the statistical similarity of predictions across sensitive groups such as race, gender, or income bracket. The fairness constraint ensures that no group systematically receives more favorable outcomes than another beyond the threshold δ, thus embedding equity into the model's structure.

To solve this constrained problem more flexibly and ensure differentiability during training, we reformulate it using the Lagrangian relaxation technique:

$$\underset{\theta}{min}L(\theta) + \lambda \cdot |\mathbb{E}_{x \sim D_{z=0}}[f_\theta(x)] - \mathbb{E}_{x \sim D_{z=1}}[f_\theta(x)]|$$

where: λ is a fairness regularization parameter that penalizes the model in proportion to the disparity between groups.

This soft-constrained formulation transforms the fairness problem into a joint optimization objective, balancing predictive accuracy and group parity through a tunable coefficient. Higher values of λ increase fairness at the potential cost of model accuracy, while lower values may privilege performance but risk reinforcing inequities. This balance is crucial for public policy applications where precision and fairness must coexist.

This fairness-aware AI model ensures that algorithmic decisions in resource-sensitive public domains do not perpetuate or exacerbate historical inequities embedded in training data. By constraining or penalizing group-level outcome disparities, the model enables equitable predictions while maintaining high performance. In practice, such fairness constraints allow AI systems to guide public institutions in healthcare triage, educational admissions, and social welfare programs with a principled commitment to justice, transparency, and inclusion. This approach transforms AI from a neutral tool into an agent of ethical governance, directly supporting the creation of inclusive and socially responsive smart cities.

Framework: Fairness-Constrained Predictive Optimization Framework (FCPOF)
Layered AI Architecture for Bias-Mitigation in Public Decision-Making

Layer	Main functionality	AI module(s) & components	Operational scope
Layer 1	Ingest and segment group-sensitive data	Group-sensitive data integration and preprocessing module	Labeled datasets with protected attributes → Prepare structured inputs for parity analysis
Layer 2	Train predictive model under fairness constraint	Fairness-constrained predictive optimization	Training data and group parity target → Optimize model with hard fairness constraints
Layer 3	Embed differentiable fairness loss	Lagrangian relaxation	Prediction loss and demographic disparity → Minimize composite fairness-augmented objective
Layer 4	Monitor group-level disparity during validation	Fairness validation and disparity monitoring module	Model outputs across subgroups → Evaluate demographic parity compliance and fairness degradation

However, as Latupeirissa et al. (2024) emphasize, these advancements are not without their challenges. Digital inclusion remains a significant issue, particularly for marginalized communities that lack access to the digital infrastructure required for effective AI systems. In many regions, limited internet access, technological literacy gaps, and unequal distribution of digital devices prevent vulnerable populations from benefiting fully from AI-powered public services.

To bridge these gaps, governments must ensure that AI equity is embedded in both policy and technology deployment. This includes investing in digital infrastructure, providing affordable access to technology, and ensuring that AI systems are developed and deployed in a manner that promotes inclusivity and fairness for all communities, regardless of their socio-economic status.

Metrics for Measuring Efficiency and Equity in AI-Driven Systems

Metrics are essential for evaluating the performance of AI-driven public service systems. Efficiency is often measured through time savings, cost reductions, and improvements in service delivery speed. For example, metrics like response times, resource utilization rates, and the accuracy of demand forecasting can all serve as indicators of efficiency in AI-driven public services (Valle-Cruz et al. 2023). To evaluate AI-driven public service systems, we introduce a multi-objective performance assessment framework that integrates both efficiency and equity indicators through a Deep Reinforcement Learning (DRL) approach with embedded fairness constraints. While traditional metrics such as response time or resource utilization offer insight into efficiency, they often overlook equity in service delivery across socio-economic groups. This framework operationalizes both dimensions by formalizing them into quantifiable reward components and loss functions within a learning environment.

We begin with the principal multi-objective policy evaluation function, which balances performance and fairness in AI decision-making:

$$\pi^* = \underset{\pi}{\arg max}\mathbb{E}_t[\alpha \cdot R_{\text{eff}}(s_t, a_t) + (1 - \alpha) \cdot R_{\text{eq}}(s_t, a_t)]$$

where, π^* is the optimal policy over time, α is a tunable weight factor that defines the prioritization between efficiency and equity (with $0 \leq a \leq 1$), $R_{\text{eff}}(s_t, a_t)$ is the efficiency reward measuring service quality and resource utilization, and $R_{\text{eq}}(s_t, a_t)$ is the equity reward capturing distributional fairness in service access across population groups.

To define efficiency in measurable terms consistent with performance indicators, we structure the efficiency reward function as follows:

$$R_{\text{eff}}(s_t, a_t) = \eta_1 \cdot (1 - \frac{T_{\text{resp}}(a_t)}{T_{\max}}) + \eta_2 \cdot U(a_t) - \eta_3 \cdot C(a_t)$$

where, $T_{\text{resp}}(a_t)$ is the system's response time after action a_t, $T_{\max}$ is the maximum tolerated response delay, $U(a_t)$ is the utility of resource utilization, $C(a_t)$ is the cost associated with action a_t, and η_1, η_2, η_3 are scalar weights adjusting the contribution of each sub-metric. This function rewards prompt, high-utility, low-cost actions, aligning with real-world efficiency indicators in AI-driven services.

To ensure that equity is evaluated in parallel, we introduce the equity reward function, which penalizes imbalance in service allocation across demographic or geographic strata:

$$R_{\text{eq}}(s_t, a_t) = -\lambda \cdot |\mu^{\text{privileged}}(a_t) - \mu^{\text{marginalized}}(a_t)|$$

where, $\mu^{\text{privileged}}(a_t)$ and $\mu^{\text{marginalized}}(a_t)$ represent the average service access levels for privileged and marginalized groups respectively under action a_t, and λ is the equity penalty coefficient. This formulation ensures that deviations in service quality across groups are detected and corrected, reinforcing distributive justice in AI policy learning.

Each component of this framework serves a specific function: the efficiency reward captures key service performance metrics—response time, utility, and cost, while the equity reward imposes a correctional force that steers learning away from discriminatory patterns. The policy is trained to converge on an optimal balance through reinforcement-learning feedback loops that simultaneously evaluate operational throughput and ethical fairness.

This AI-driven framework not only embeds real-world performance indicators but also provides a mathematically grounded method to integrate them with equity assessments. By incorporating efficiency and fairness into the reward structure, the model ensures that public service AI systems achieve measurable performance gains while maintaining ethical obligations to underserved populations. This dual-focus evaluation architecture allows public administrators and urban planners to assess algorithmic decisions on both operational and social grounds, offering a robust and transparent foundation for the future of equitable smart governance.

Framework: Multi-Objective Performance Evaluation Framework (MOPEF)
Layered AI Architecture for Efficiency and Equity in Public Service Evaluation

Layer	Main functionality	AI module(s) & components	Operational scope
Layer 1	Gather public service performance and demographic data	Performance data integration and preprocessing module	Service logs, access data, demographic indicators → Prepare structured inputs for evaluation
Layer 2	Optimize policy with multi-objective reward function	Scalarized deep reinforcement learning (DRL)	Efficiency and equity rewards → Learn policy balancing performance and fairness
Layer 3	Calculate efficiency reward	Performance-based reward function	Response time, utility, and cost metrics → Compute real-time service efficiency
Layer 4	Apply fairness-aware equity reward	Demographic parity penalty function	Service access gaps across groups → Penalize outcome disparities in decision-making

(continued)

(continued)

Layer	Main functionality	AI module(s) & components	Operational scope
Layer 5	Monitor policy impacts across population groups	Policy impact monitoring and equity evaluation module	Policy outcomes and group-level statistics → Evaluate performance and fairness jointly over time

On the other hand, equity in AI-driven public services can be measured through various metrics that assess the distribution of services across socio-economic groups, geographic regions, and demographic categories. These metrics help ensure that AI systems do not disproportionately benefit certain groups at the expense of others. For instance, measuring the distribution of healthcare resources across urban and rural areas or analyzing access to public services by different income levels can highlight imbalances in service provision. Latupeirissa et al. (2024) stress the importance of conducting regular equity assessments to monitor service accessibility and ensure that AI systems are not exacerbating disparities. One key tool in this process is the use of equity audits, which assess whether the outcomes of AI-driven services align with the goal of equitable access for all demographic groups. These audits identify disparities in service outcomes and provide insights into how AI-driven decisions can be adjusted to improve fairness in distribution.

Such equity audits are crucial for maintaining transparency and accountability in AI governance. They allow policymakers and organizations to evaluate whether AI systems are genuinely serving the public interest, ensuring that the technological transformation driven by AI leads to positive, inclusive outcomes across society.

To operationalize equity audits in AI-driven public services, we introduce a fairness evaluation model based on a group benefit disparity function. This function quantifies disparities in service outcomes across demographic groups and enables public institutions to assess whether the AI system is allocating resources or opportunities fairly. It is particularly relevant in domains such as healthcare delivery, welfare access, and digital service deployment, where outcome equity is critical to maintaining social legitimacy and institutional trust.

At the core of this model lies the expected benefit gap function across two demographic groups:

$$\Delta_{eq} = |\mathbb{E}_{x \sim D_A}[f(x)] - \mathbb{E}_{x \sim D_B}[f(x)]|$$

where: Δ_{eq} is the equity disparity metric, $\mathbb{E}_{x \sim D_A}[f(x)]$ and $\mathbb{E}_{x \sim D_B}[f(x)]$ are the expected outcomes of the AI system for individuals drawn from groups A and B, respectively (e.g., rural vs. urban residents, high-income vs. low-income users), and $f(x)$ is the model's predicted benefit or decision outcome (e.g., eligibility score, treatment allocation, access level). This function measures the absolute difference in predicted benefits received by different subgroups and forms the quantitative backbone of an equity audit. A larger Δ_{eq} value indicates a wider disparity in access or outcomes between demographic groups, thereby signaling potential bias or structural inequality within the AI system.

To account for multi-group scenarios and population proportions, we extend the function using a population-weighted generalization:

$$\Delta_{pop} = \sum_{g=1}^{G} w_g \left|\mathbb{E}_{x\sim D_g}[f(x)] - \mathbb{E}_{x\sim D}[f(x)]\right|$$

where: Δ_{pop} is the population-weighted disparity score across all G groups, w_g is the relative population weight of group g, $\mathbb{E}_{x\sim D_g}[f(x)]$ is the group-specific expected outcome, and $\mathbb{E}_{x\sim D}[f(x)]|$ is the overall population average.

This formulation ensures that the model fairly evaluates AI performance across all groups, not just binary comparisons, and penalizes disproportionate disparities that affect minority or underrepresented populations. It serves as a diagnostic mechanism for identifying which subgroups are consistently underserved and where adjustments in the AI system or resource distribution may be necessary.

This equity-focused AI model enables precise auditing of outcome disparities in AI-driven public services. By quantifying the distribution of benefits or decisions across socio-economic and demographic groups, it empowers policymakers to detect and respond to fairness violations. The population-weighted disparity score introduces statistical rigor into equity audits, ensuring that observed gaps are not only identified but also contextualized within the broader social structure. This model supports transparent AI governance by offering a replicable, mathematically grounded audit trail for how AI decisions impact diverse communities, ensuring that public AI systems are accountable, inclusive, and equitably aligned with societal needs.

Framework: Equity Audit Model for AI-Driven Public Services
Layered AI Architecture for Group Fairness Evaluation and Disparity Auditing

Layer	Main functionality	AI module(s) & components	Operational scope
Layer 1	Collect group-specific outcome data	Demographic outcome data integration module	Service results and demographic attributes → Organize by socio-economic groups
Layer 2	Measure group benefit disparity	Group benefit disparity function	Outcome expectations for two groups → Quantify equity gap
Layer 3	Compute population-weighted disparity	Population disparity audit module	Multiple group weights and expectations → Assess equity across full population
Layer 4	Generate equity audit report	Equity reporting and fairness adjustment module	Disparity metrics → Summarize gaps and suggest fairness adjustments

Figure 2.1.5 presents key strategies for leveraging AI to enhance efficiency and equity in public service delivery. It focuses on three primary areas: optimizing resource allocation, ensuring equitable access to services, and using metrics to measure performance. AI optimizes resource allocation by predicting service demand and enabling real-time adjustments, which improves resource distribution in areas

Fig. 2.1.5 AI-driven efficiency and equity in public service delivery

like healthcare and emergency response. To ensure equity, AI bridges urban–rural gaps, reduces biases in resource distribution, and addresses digital inclusion challenges, promoting fair access to services. Metrics assess efficiency through indicators like cost savings and response times, while equity is evaluated through service distribution and equity audits, ensuring alignment with equitable access goals. Each area is represented with color-coded nodes and sub-nodes, providing a clear overview of AI's role in public service transformation.

AI has the potential to revolutionize public service delivery by optimizing resource allocation and ensuring equitable access. The ability of AI to predict demand and dynamically allocate resources significantly improves efficiency, while its capacity to mitigate biases ensures fairness. However, challenges related to digital inclusion and access must be addressed to realize the full potential of AI-driven public services.

2.1.2 AI and Equity in Public Service Delivery

AI technologies are increasingly being used in public service delivery to address disparities among different demographic groups. This application of AI holds promise in promoting equitable service delivery by tailoring services to individual needs and identifying previously underserved populations. AI has the potential to analyze vast datasets, enabling governments to detect patterns of inequality and develop more targeted interventions to reduce them (O'Brien et al. 2022). As O'Brien et al. (2022) emphasize, the strategic use of AI in this context can transform service delivery from reactive to proactive, achieving measurable progress toward social equity in cities of the future.

To support equitable public service delivery, we propose a high-precision AI-driven analytical model based on extreme gradient boosting (XGBoost). This model is particularly well-suited to analyze large-scale, heterogeneous demographic datasets and detect structural disparities in access patterns. Its capacity to process non-linear relationships and assign differential importance to features makes it ideal for identifying underserved populations and tailoring interventions accordingly.

The AI-driven disparity detection model using XGBoost is defined by its additive learning structure:

$$\hat{y}_i = \sum_{t=1}^{T} f_t(x_i), f_t \in \mathcal{F}$$

where, $\hat{y}_i$ is the predicted service adequacy score for individual i, $f_t(x_i)$ is the predictive function of the t-th decision tree based on input features x_i, T is the total number of trees, and $\mathcal{F}$ is the space of regression trees used to model interactions across demographic variables such as age, location, income, or disability status.

To ensure that the learned model remains interpretable and avoids overfitting—especially important in sensitive applications like equity assessments—we include a complexity penalty in the overall objective function. The XGBoost objective function with regularization is:

$$\mathcal{L}(\phi) = \sum_{i=1}^{n} \ell(\hat{y}_i, y_i) + \sum_{t=1}^{T} \Omega(f_t)$$

where, $\ell(\hat{y}_i, y_i)$ is the loss function measuring the difference between the predicted and actual service equity score for individual i, and $\Omega(f_t)$ is the regularization term that penalizes the complexity of each decision tree to ensure model simplicity and generalizability.

The regularization component $\Omega(f_t)$ is defined as:

$$\Omega(f_t) = \gamma \cdot |L_t| + \frac{1}{2}\lambda \sum_{j \in L_t} w_j^2$$

where, $|L_t|$ is the number of leaves in tree t, w_j is the weight assigned to the j-th leaf, γ controls the penalty for model complexity (number of leaves), and λ penalizes large leaf weights to avoid over-reliance on any single pattern in the data.

This formulation allows the XGBoost algorithm to evaluate how different population groups access public services and predict which individuals or communities are likely to experience insufficient access. Governments can then use these predictions to implement preemptive or corrective policies, such as mobile healthcare units for rural populations, or language-accessible digital platforms for immigrant communities.

This XGBoost-driven disparity detection model enables governments to move beyond anecdotal or retrospective equity assessments. By learning patterns from multidimensional demographic data, the model offers real-time insights into service gaps and allows for precision targeting of public resources. Its mathematical formulation ensures both high accuracy and interpretability, critical for transparent governance.

Framework: XGBoost-Driven Public Equity Assessment Framework (XPEAF)
Layered AI Architecture for Fairness-Driven Public Service Targeting

Layer	Main functionality	AI module(s) & components	Operational scope (input → processing → output)
Layer 1	Contextual mapping of demographic inequities	Demographic data integration and context encoding module	Demographic and service access data → Categorization and context tagging → Preprocessed equity landscape
Layer 2	Disparity detection and service adequacy scoring	Extreme gradient boosting (XGBoost)	Preprocessed equity landscape → Tree-based disparity learning → Predicted adequacy scores per individual
Layer 3	Regularized model learning for equity assessment	XGBoost regularized objective function	Predicted adequacy scores and true values → Loss minimization with regularization → Generalized fair decision boundaries
Layer 4	Model interpretability and feature attribution	SHAP feature attribution integrated with XGBoost	Trained XGBoost model → Feature importance estimation → Local and global interpretability reports
Layer 5	Policy deployment and equity monitoring	Policy implementation and fairness monitoring module	Equity reports and flagged disparities → Human evaluation and policy enforcement → Real-time service adjustment and fairness tracking

AI-driven platforms have been instrumental in improving access to services, particularly in areas like healthcare and social welfare, where historically marginalized groups, such as ethnic minorities, have been disproportionately underserved. For example, AI tools can identify geographic areas with limited access to healthcare facilities and prioritize the allocation of resources accordingly (O'Brien et al. 2022). Furthermore, AI can be deployed to ensure that public services are accessible to people with disabilities by offering tailored communication methods and assistive technologies (Agarwal et al. 2023). To address the needs of historically marginalized populations and individuals with disabilities, we introduce a geospatial fair resource allocation model using deep reinforcement learning (DRL) with proximal policy optimization (PPO). This framework allows for dynamic learning of optimal service delivery actions in heterogeneous spatial environments while ensuring equitable distribution through embedded fairness metrics. The model adapts in real-time to changing service demand by learning which geographic zones are consistently underserved and allocating resources proportionally. The system maintains continuous feedback using health, demographic, and disability-access datasets, optimizing its decisions across time and space.

The principal DRL objective function, integrating spatial fairness into policy learning, is formulated as:

$$\pi_\theta^* = \underset{\theta}{\arg max}\mathbb{E}_t[\frac{\pi_\theta(a_t|s_t)}{\pi_{\theta_{\text{old}}}(a_t|s_t)} \cdot \hat{A}_t - \beta \cdot \mathcal{L}_{\text{geo-fair}}(s_t, a_t)]$$

where, $(a_t|s_t)$ is the updated policy selecting action a_t in state s_t, $\pi_{\theta_{\text{old}}}(a_t|s_t)$ is the old policy for PPO stability, $\hat{A}_t$ is the advantage estimate assessing the value of action a_t, β is a fairness regularization coefficient, and $\mathcal{L}_{\text{geo-fair}}(s_t, a_t)$ is the geospatial fairness loss penalizing unequal service access across regions.

To explicitly capture spatial inequities such as ethnic minorities or individuals with disabilities experiencing lower service accessibility we define the geospatial fairness loss function as:

$$\mathcal{L}_{\text{geo-fair}}(s_t, a_t) = \sum_{k=1}^{K} |\mu_k(a_t) - \overline{\mu}(a_t)|$$

where, $\mu_k(a_t)$ is the average service accessibility level in region k after action a_t, $\overline{\mu}(a_t)$ is the mean access level across all K regions. This function imposes a penalty proportional to deviation from geographic equality, ensuring that underserved areas (rural, minority-dense, or disability-impacted) are prioritized during allocation.

To evaluate immediate performance outcomes from each allocation decision, we apply a social utility-based reward function:

$$R(s_t, a_t) = U^{\text{health}}(a_t) + U^{\text{disability}}(a_t) - C(a_t)$$

where, $U^{\text{health}}(a_t)$ measures the utility gained from improved healthcare access, $U^{\text{disability}}(a_t)$ reflects utility from assistive or accessible service deployment for people with disabilities, and $C(a_t)$ is the cost of executing action a_t. This reward design promotes decisions that benefit both underserved ethnic regions and disability groups, while minimizing resource expenditure.

This DRL model enables real-time learning of optimal strategies to deploy healthcare and social welfare services equitably across geographic zones. Through PPO-based policy refinement and spatial fairness constraints, the system continuously corrects access gaps. The incorporation of disability-access rewards and health utility ensures that tailored communication methods and assistive technologies are not peripheral features but mathematically embedded into the core learning objective. Such AI-driven approaches provide a scientifically grounded mechanism to ensure that public services no longer reinforce structural exclusion, but instead prioritize those who have historically been marginalized.

Framework: DRL-Based Geospatial Fairness Allocation Framework (DRL-GFAF)
Layered AI Architecture for Equitable Public Service Distribution

Layer	Main functionality	AI module(s) & components	Operational scope (input → processing → output)
Layer 1	Geospatial contextual mapping and demand identification	Geospatial data integration and demand mapping module	Healthcare, disability, and geographic data → Underserved zone detection → Spatial service demand map
Layer 2	Equity-aware policy learning using PPO	Proximal policy optimization (PPO)	State-action pairs in underserved zones → Policy training with PPO → Equity-aware allocation policy
Layer 3	Geospatial fairness penalty integration	Geospatial fairness loss function	Policy actions and access metrics → Fairness penalty computation → Balanced access optimization
Layer 4	Composite social utility-based reward evaluation	Social utility-based reward function	Policy actions and outcome data → Reward evaluation using health and disability metrics → Utility score for policy refinement
Layer 5	Real-time policy deployment and adaptive monitoring	Adaptive policy deployment and real-time monitoring module	Learned equity policies and live feedback → Dynamic deployment and monitoring → Updated resource allocation decisions

However, despite these advancements, challenges remain. One major issue is algorithmic bias, where AI systems unintentionally perpetuate existing inequalities through biased data inputs. For example, if an AI model is trained on historical data that reflects systemic biases (such as lower service access for certain ethnic groups), it may reinforce these patterns rather than disrupt them (Agarwal et al. 2023). This is particularly concerning in healthcare, where AI algorithms have been shown to allocate fewer resources to Black patients due to bias in the data used to train these systems (Agarwal et al. 2023). To mitigate the systemic bias perpetuated by historical data in sensitive domains such as healthcare, we propose an AI-driven fairness framework based on adversarial debiasing. This approach introduces a two-player learning dynamic in which the primary predictor aims to model the target variable accurately, while an adversary is simultaneously trained to detect sensitive attributes (e.g., race, gender) from the predictor's outputs. If the adversary is unable to distinguish protected group membership, it implies that the learned representation is fair with respect to that attribute.

At the core of this framework lies the minimax optimization objective:

$$\underset{\theta_f}{min}\underset{\theta_a}{max}\mathcal{L}_f(f_{\theta_f}(x), y) - \lambda \cdot \mathcal{L}_a(a_{\theta_a}(f_{\theta_f}(x)), z)$$

where: $\mathcal{L}_f$ is the prediction loss (e.g., cross-entropy or MSE) for the main task with model f_θ, $\mathcal{L}_a$ is the adversary's loss in predicting the protected attribute z, λ is a trade-off parameter that governs the balance between accuracy and fairness, x is the input feature vector, y is the true label, and a_{θ_a} is the adversarial network trying to predict the sensitive attribute from the model's output. The minimization step encourages accurate predictions of the target variable y, while the adversarial maximization penalizes any encodings in the prediction that allow inference of the protected variable z. The adversary's inability to predict z implies statistical independence between group membership and outcome, enforcing group fairness.

To improve training stability and support fairness across multiple demographic attributes, the adversarial signal can be projected back to the predictor through gradient reversal:

$$\nabla_{\theta_f} \leftarrow \nabla_{\theta_f}\mathcal{L}_f - \lambda \cdot \nabla_{\theta_f}\mathcal{L}_a$$

where: ∇_{θ_f} denotes the gradient updates to the predictor's parameters, with the adversarial loss subtracted to discourage encoding bias-related features in $f_{\theta_f}(x)$.

This gradient adjustment mechanism ensures that the primary model learns not only to be accurate but also to actively remove bias-inducing patterns from its internal representations. In healthcare, this helps prevent AI systems from implicitly favoring dominant population groups, even when trained on historical datasets that embed structural inequalities.

The adversarial debiasing framework allows AI models to learn representations that are both accurate and fair by simultaneously optimizing prediction performance and group fairness. By neutralizing the influence of sensitive variables during training, this model prevents the reinforcement of historical biases, especially in socially sensitive applications such as healthcare allocation. It provides a robust defense against algorithmic discrimination by mathematically enforcing fairness through adversarial dynamics. This model is particularly valuable for public sector deployments of AI, where ethical compliance and social equity are foundational requirements for responsible technological governance.

Framework: Adversarial Debiasing Fairness Framework (ADFF)

Layered AI Architecture for Bias Mitigation in Sensitive Public Sector Applications

Layer	Main functionality	AI module(s) & components	Operational scope (input → processing → output)
Layer 1	Bias risk identification from historical training data	Bias detection and data audit module	Historical training datasets with known bias → Pre-audit of sensitive group disparities → Flagged attributes for fairness-sensitive learning
Layer 2	Fair representation learning via adversarial debiasing	Adversarial debiasing minimax module	Input features and sensitive attributes → Minimax adversarial training → Group-fair encoded representations

(continued)

(continued)

Layer	Main functionality	AI module(s) & components	Operational scope (input → processing → output)
Layer 3	Gradient reversal for bias suppression in backpropagation	Gradient reversal layer	Predictor and adversary gradient signals → Bias suppression via reversed gradient → Bias-reduced parameter updates
Layer 4	Ethical evaluation and deployment for sensitive domains	Ethical deployment and governance review module	Group-fair model outputs → Ethical review and policy criteria → Responsible model deployment in healthcare

Challenges and Barriers to Achieving Equity

One of the primary barriers to achieving equity in AI-driven public service delivery is the quality of the data used to train AI systems. Biases in training data, whether arising from socio-economic factors, racial disparities, or historical systemic discrimination, can significantly affect the outcomes of AI algorithms. As Agarwal et al. (2023) highlight, AI systems are only as unbiased as the data they learn from. If the data reflects historical inequities, the AI systems are likely to replicate these disparities in their predictions and decision-making processes, perpetuating existing social injustices.

To mitigate historical inequities embedded in biased training data, we propose an AI-driven framework based on fair representation learning. This approach transforms the input data into a latent representation that is simultaneously informative for prediction while statistically independent of sensitive attributes such as race, gender, or income. It is particularly suitable for high-stakes public applications like healthcare, housing, or employment allocation, where fairness in the input representation is critical for equitable outcomes.

The core objective is to learn a transformation function $\phi(x)$ that maps raw input data x into a fair latent space h, by minimizing a loss function that balances three goals: predictive accuracy, data reconstruction, and independence from the sensitive attribute z:

$$\min_{\theta_\phi, \theta_y, \theta_x} \mathcal{L}_y(y, \hat{y}) + \beta \cdot \mathcal{L}_x(x, \hat{x}) + \gamma \cdot I(h; z)$$

where: $\mathcal{L}_y(y, \hat{y})$ is the prediction loss between true labels y and predicted outputs $\hat{y} = f_{\theta_y}(h)$, $\mathcal{L}_x(x, \hat{x})$ is the reconstruction loss between input x and its reconstruction $\hat{x} = g_{\theta_z}(h)$, $h = \phi_{\theta_\phi}(x)$, is the learned fair representation, z is the sensitive attribute, $I(h; z)$ is the mutual information between the representation and the sensitive attribute, andβ, γ are regularization weights controlling the trade-off among the three objectives.

This function promotes the learning of a feature space that retains relevant information for task performance ($\mathcal{L}_y$), preserves data integrity ($\mathcal{L}_x$), and minimizes leakage of sensitive group information ($I(h; z)$), thus reducing the risk that AI decisions are indirectly influenced by protected attributes.

To approximate and optimize the mutual information term $I(h; z)$, an adversarial regularization strategy is often employed:

$$I(h; z) \approx \max_{\theta_d} \mathbb{E}_{(h,z)}[\log d_{\theta_d}(z|h)]$$

where: $d_{\theta_d}(z|h)$ is an adversarial discriminator trained to predict the sensitive attribute z from the latent representation h, and θ_d are its parameters. Minimizing the adversary's performance enforces independence between the representation and the sensitive attribute.

This fair representation model enables AI systems to neutralize the effect of historical data bias by learning task-relevant features that are independent of sensitive attributes. It ensures that predictions are made based on meritocratic signals rather than inherited structural inequities, thus promoting fairness in public sector AI applications. In doing so, it addresses the foundational barrier of biased data quality and offers a mathematically sound pathway for building equitable decision-making systems that align with the principles of social justice and responsible AI governance.

Framework: Fair Representation Learning Framework (FRLF)
Layered AI Architecture for Fair Latent Representations in Public Decision Systems

layer	Main functionality	AI module(s) & components	Operational scope (input → processing → output)
Layer 1	Bias risk identification in training data	Data bias detection and pre-audit module	Biased dataset → Pre-processing audit → Sensitive attributes flagged
Layer 2	Fair representation learning	Fair representation learning module	Raw data → Representation learning → Task-ready latent space
Layer 3	Mutual information adversarial debiasing	Mutual information adversarial discriminator	Representation + sensitive data → Adversarial MI minimization → Independent latent embeddings
Layer 4	Ethical evaluation and deployment in public services	Ethical deployment and fairness compliance module	Fair latent model → Ethical review → Deployment in high-stakes application

In addition to biased data, another significant barrier is the digital divide. The unequal access to digital services and technologies remains a critical issue across different socio-economic groups. Low-income individuals, ethnic minorities, and rural populations are disproportionately affected by the lack of access to essential technologies, such as high-speed internet, smartphones, or other devices necessary to engage with AI-driven services (O'Brien et al. 2022). This divide deepens existing social inequalities, as those who would benefit most from AI-enhanced services are often the least equipped to access them. As a result, the digital divide exacerbates the inequitable distribution of public resources, preventing the full potential of AI from being realized in underserved communities.

In tackling the digital divide, we first need to establish a mechanism for detecting accessibility risks across geographies and populations. The foundation of such detection is built upon correlating infrastructure availability (e.g., broadband access, device penetration) with socio-demographic variables (e.g., income, ethnicity, education). To formalize this correlation within a learnable system, we construct a digital divide risk function, where the objective is to model the probability that a given unit (individual or region) lacks sufficient digital access.

We define the accessibility risk model as follows:

$$\hat{r}_i = \sigma\left(w^\top x_i + b\right)$$

where: $\hat{r}_i$ is the predicted digital exclusion risk for instance i, x_i is the demographic-infrastructure feature vector, w is the learned weight vector, b is the bias term, $\sigma(\cdot)$ is the sigmoid activation function. This function transforms demographic-infrastructure profiles into exclusion probability estimates, allowing the model to flag regions and groups with high vulnerability to digital inaccessibility.

Once high-risk areas are detected, the next phase involves optimally allocating AI outreach resources to mitigate these gaps. For this, we employ a reinforcement learning-based equity-aware resource allocation model that iteratively refines policy decisions based on exclusion risk signals and real-time service feedback. The policy is optimized using a proximal policy optimization (PPO) variant, guided by an equity-weighted reward function.

The clipped surrogate objective guiding the learning of this policy is:

$$\mathcal{L}^{\text{PPO}}(\theta) = \mathbb{E}_t[min(r_t(\theta)A_t, \text{clip}(r_t(\theta), 1-\epsilon, 1+\epsilon)A_t)]$$

where: $\mathcal{L}^{\text{PPO}}(\theta)$ is the PPO loss function to be minimized, $r_t(\theta) = \frac{\pi_\theta(a_t|s_t)}{\pi_{\theta_{\text{old}}}(a_t|s_t)}$ is the probability ratio of the new and old policies, A_t is the advantage function at timestep t, ϵ is the clipping parameter controlling how much policy deviation is allowed. This function refines the allocation policy to ensure that service outreach is optimized under fairness-preserving constraints, particularly for digitally marginalized populations.

To close the loop between service allocation and real-world access constraints, we must consider the technical feasibility of delivering AI services in bandwidth-limited settings. Therefore, we define a bandwidth-constrained delivery optimization mechanism, where AI models are compressed and selectively distributed based on regional connectivity.

Let the delivery compression objective be formalized by minimizing a latency- and utility-aware deployment cost:

$$\min_{\phi} \mathbb{E}_j[\alpha \cdot \text{Latency}_j(\phi) + \beta \cdot (1 - \text{Utility}_j(\phi))]$$

where: ϕ denotes the compression strategy for model deployment, $\text{latency}_j(\phi)$ is the estimated latency for deploying the model ϕ in region j, $\text{utility}_j(\phi)$ is the predicted utility score of the compressed model for that region, α, β are trade-off parameters. This function ensures AI model delivery is both latency-efficient and utility-preserving, especially in digitally underserved zones where high-bandwidth services are not feasible.

This integrated architecture provides a principled and implementable framework for bridging the digital divide. The logical flow from identifying exclusion risk, to equity-aware outreach, to feasibility-driven deployment ensures that AI systems do not merely encode fairness at the algorithmic level but also operationalize it across access and infrastructure boundaries.

Framework: AI Accessibility Equity Framework (AAEF)
Layered AI Architecture for Overcoming the Digital Divide in Public Service Access

Layer	Main functionality	AI module(s) & components	Operational scope (input → processing → output)
Layer 1	Digital infrastructure assessment and mapping	Digital infrastructure data integration and gap mapping module	Geospatial infrastructure, internet access, device ownership → Mapping of access gaps → Regional digital infrastructure profiles
Layer 2	Accessibility risk detection in underserved populations	Digital divide detection module using demographic-infrastructure correlation	Infrastructure profiles + socio-economic data → Risk modeling of digital exclusion → Identified high-vulnerability groups
Layer 3	Equity-aware AI service outreach planning	Equity-Aware Resource Allocation Via Reinforcement Learning	Exclusion risk profiles + service availability → Outreach policy optimization → Tailored resource deployment strategies
Layer 4	Adaptive low-bandwidth AI delivery optimization	Bandwidth-aware AI model compression and distribution	AI models + regional bandwidth constraints → Compression and delivery tuning → Optimized deployment in limited-access areas
Layer 5	Monitoring and feedback for inclusive service coverage	Inclusive service monitoring and equity feedback module	Service utilization logs and feedback loops → Inclusion performance assessment → Adjusted equity-focused outreach

Finally, the transparency of AI systems is another significant challenge. Many AI models operate as "black boxes," where it is difficult to understand how decisions are made. This lack of transparency can lead to public mistrust, particularly if AI systems are perceived as making unfair or biased decisions (Agarwal et al. 2023). Ensuring that AI systems are both transparent and accountable is critical for gaining public trust and promoting equity in public service delivery.

AI offers significant potential for improving equity in public service delivery by optimizing resource allocation and identifying underserved populations. However, challenges such as algorithmic bias, the digital divide, and the lack of transparency must be addressed to fully realize this potential. Policymakers and technologists must work together to ensure that AI systems are designed and implemented in ways that promote fairness, transparency, and accountability.

To overcome the opacity of complex AI systems and promote trust and accountability in public service delivery, we propose the integration of a mathematically grounded interpretability model known as SHapley Additive exPlanations (SHAP). This framework attributes the prediction of a model to each input feature using principles from cooperative game theory, where features are treated as "players" contributing to the final decision. SHAP values provide a consistent and locally accurate explanation of how individual model inputs influence specific outputs transforming opaque models into auditable systems.

At the core of this model lies the Shapley value decomposition of a model's output:

$$f(x) = \phi_0 + \sum_{i=1}^{M} \phi_i$$

where: $f(x)$ is the original model's prediction for input x, ϕ_0 is the expected prediction (baseline value) across all data instances, M is the number of input features, and ϕ_i is the Shapley value that quantifies the contribution of feature ii to the deviation from the baseline prediction.

The Shapley value ϕ_i for each feature is computed by evaluating its marginal contribution over all possible feature subsets:

$$\phi_i = \sum_{S \subseteq N \setminus \{i\}} \frac{|S|!(M-|S|-1)!}{M!} [f_{S \cup \{i\}}(x_{S \cup \{i\}}) - f_S(x_S)]$$

where: S is a subset of features excluding i, N is the full feature set, $f_S(x_S)$ is the model trained with only features in S, and $f_{S \cup \{i\}}(x_{S \cup \{i\}})$ is the model with feature i added. The factorial-based weighting ensures a fair average over all feature orderings, aligning with game-theoretic fairness. This formulation ensures that every feature's contribution is fairly attributed, regardless of model complexity. When applied to AI systems in public domains such as healthcare prioritization or resource allocation SHAP enables stakeholders to audit not just outputs, but the specific logic and data relationships that lead to decisions.

The SHAP framework provides a method for explaining the predictions of any machine learning model, including complex ensembles and deep neural networks. By transforming black-box outputs into transparent, human-interpretable explanations, SHAP supports accountability and public trust in AI-driven decisions. Its ability to isolate how specific features such as income, location, or ethnicity influence model behavior is crucial for auditing fairness and ensuring that algorithmic decisions align with social equity goals. In the context of public service delivery, SHAP not only enhances technical transparency but also enables ethically responsible AI deployment that respects democratic governance and civic oversight.

Framework: Transparent AI Explanation Framework Using SHAP (TAEF-SHAP)
Layered AI Architecture for Interpretability and Accountability in Public Sector Models

Layer	Main functionality	AI module(s) & components	Operational scope (input → processing → output)
Layer 1	Audit requirement identification for public sector AI Systems	Model audit identification and risk screening module	AI models used in public decision systems → Identification of opaque or high-risk models → Flagged candidates for explanation
Layer 2	Prediction attribution using additive SHAP explanation	SHAP additive attribution model	Selected AI model + input data → SHAP-based additive feature attribution → Local explanation per decision instance
Layer 3	Shapley value computation for feature-level transparency	Shapley value decomposition module	AI model + feature subsets → Shapley value aggregation across permutations → Quantified feature contributions
Layer 4	Public accountability and fairness certification mechanism	Transparency certification and public accountability module	SHAP outputs + stakeholder review → Fairness and transparency audit → Documentation and public accountability

2.2 Conclusion

This chapter demonstrates that the integration of artificial intelligence into public services is not a marginal reform but a structural redefinition of governance itself. By embedding advanced learning systems into the allocation, evaluation, and oversight of resources, governments can transcend the inefficiencies of reactive administration and move toward proactive, equitable, and adaptive service delivery. The proposed frameworks from multi-sector optimization models and contextual bandit strategies to fairness-aware reinforcement learning and interpretability tools collectively reveal how AI can institutionalize efficiency without sacrificing justice.

The analysis shows that AI's power lies in its ability to unify seemingly conflicting imperatives: speed with deliberation, efficiency with fairness, and autonomy with accountability. Through mathematically rigorous formulations, the chapter illustrates how predictive accuracy, resource optimization, and bias mitigation can be harmonized within one coherent architecture. More than a technical achievement, this balance is presented as a democratic necessity ensuring that automation does not deepen existing divides but instead acts as a corrective to structural inequities.

Equally, the chapter does not neglect the barriers. Bias in training data, the digital divide, and the opacity of algorithms remain formidable challenges. Yet through fairness-constrained optimization, equitable reinforcement learning, and transparent interpretability mechanisms, these risks can be systematically addressed. What emerges is a vision of governance where AI becomes not merely a tool of efficiency but a guarantor of legitimacy, inclusivity, and resilience.

The chapter positions AI-driven public services as the foundation of a new social contract: one where governments are empowered to deliver faster, fairer, more transparent outcomes, and citizens can trust that technological progress serves the collective good.

References

Agarwal, R., Bjarnadottir, M., Rhue, L., Dugas, M., Crowley, K., & Clark, J. (2023). Addressing algorithmic bias and the perpetuation of health inequities: An AI bias-aware framework. *Health Policy and Technology*. https://doi.org/10.1016/j.hlpt.2022.100702

Latupeirissa, J. J. P., Dewi, N. L. Y., Prayana, I. K. R., Srikandi, M. B., Ramadiansyah, S. A., & Pramana, I. B. G. A. Y. (2024). Transforming public service delivery: A comprehensive review of digitization initiatives. *Sustainability*. https://doi.org/10.3390/su16072818

O'Brien, N., Van Dael, J., Clarke, J., Gardner, C., O'Shaughnessy, J., Darzi, A., & Ghafur, S. (2022). Addressing racial and ethnic inequity in data-driven health technologies. *Institute of Global Health Innovation*. https://ora.ox.ac.uk/objects/uuid:76cf09cf-ddf7-43ae-a724-9b447136356a

Osborne, S. P., Cucciniello, M., Nasi, G., & Zhu, E. (2022). Digital transformation, artificial intelligence, and effective public services: Challenges and opportunities. *Global Public Policy and Governance*. https://doi.org/10.1007/s43508-022-00058-7

Singh, V., & Chobotaru, J. (2022). Digital divide: Barriers to accessing online government services in Canada. *Administrative Sciences*. https://doi.org/10.3390/admsci12030112

Valle-Cruz, D., Garcia-Contreras, R., & Contreras, D. (2023). Towards AI-driven transformation and smart data management: Emerging technological change in the public sector. *Public Policy and Administration*. https://doi.org/10.1177/09520767231188401

Chapter 3
Artificial Intelligence, Ethics, and Governance: Safeguarding Privacy, Equity, and Public Trust

This chapter establishes the ethical and policy foundations for governing artificial intelligence in high-stakes public domains. It confronts the core tension of our era: how to harness AI's analytic speed and scale without eroding privacy, equity, and democratic legitimacy. We move beyond abstract principles to operational architectures, showing how rights become code and policy becomes an enforceable design constraint across the full AI lifecycle data collection, model training, deployment, and oversight.

First, we delineate the ethical perimeter: fairness, transparency, accountability, and human agency in settings where decisions carry moral weight healthcare, justice, and public administration. We then translate these values into implementable mechanisms. Differential privacy and its training-time realization via DP-SGD secure individuals against re-identification in data-intensive urban systems, while privacy-by-design reframes protection as a precondition rather than a retrofit. Regulation-aware federated learning embeds consent, data minimization, and legal compliance into decentralized model training, making large-scale learning possible without centralizing personal data.

Second, we address legitimacy at the point of decision. Group-level disparity metrics and surveillance-exposure audits give administrators a way to detect and correct structural harms as they emerge. Compliance dashboards and transparency reports transform black-box pipelines into systems the public can interrogate, contest, and improve. Throughout, we insist that human judgment remains sovereign: AI should augment professional discretion, not replace it.

The result is a governance blueprint technically precise and institutionally practical that allows states, cities, and public agencies to build AI that is fast and fair, scalable and rights-preserving. Readers will find concrete frameworks, design patterns, and accountability controls that convert ethical intent into measurable, auditable guarantees. The chapter's wager is clear: encode privacy, equity, and accountability into the architecture now or risk encoding them out of the society we are building.

D. Sargiotis, *Empathy in Algorithms*, https://doi.org/10.1007/978-3-032-32128-2_3

3.1 Ethical Implications of AI in Public Policy

The relationship between ethics and AI is deeply intertwined, influencing the success or failure of AI deployment across diverse sectors. At its core, the ethical dimensions of AI revolve around ensuring fairness, transparency, accountability, and minimizing potential harm. AI's capacity to autonomously analyze data and make decisions raises fundamental concerns about bias, the opacity of algorithms, and the degree of human oversight necessary in the decision-making process. These ethical questions are especially salient in high-impact domains, such as healthcare, criminal justice, and public services, where the consequences of biased or opaque algorithms can have far-reaching effects on people's lives (Vesnic-Alujevic et al. 2020).

AI presents several ethical challenges, particularly in the risk of exacerbating inequality and discrimination. When AI systems are trained on biased or incomplete data, they tend to replicate and, in some cases, amplify existing societal disparities. This phenomenon has been observed in applications such as predictive policing and risk assessments in the criminal justice system, where AI systems have disproportionately impacted vulnerable groups, reinforcing cycles of inequality (Council of Europe 2018). Consequently, ethical considerations emphasize the urgent need for transparent mechanisms, such as comprehensive audit trails, that enable public scrutiny of AI decision-making processes. Furthermore, there is a call for rigorous efforts to train AI systems on balanced, diverse, and representative datasets, ensuring that these technologies do not perpetuate or amplify biases, but rather promote fairness and inclusivity (Pflanzer et al. 2023).

3.2 Governance and Accountability

Accountability is a cornerstone of ethical AI usage, yet it remains one of the most complex issues in AI governance. The question of who is responsible for AI-driven decisions, especially when these decisions result in harm, requires careful consideration. Should accountability rest with the designers of the AI, the data providers, or the end-users who deploy these systems? This question becomes particularly urgent when AI systems operate autonomously, as seen in high-stakes fields like healthcare or autonomous vehicles (Pflanzer et al. 2023). The delegation of responsibility is further complicated when the decision-making process is opaque, and the actions of AI systems are difficult to predict or control. As AI technology continues to evolve, governance structures must adapt by developing regulatory frameworks that can keep pace with these advancements, ensuring that accountability is clearly defined and enforced.

Governments worldwide are beginning to acknowledge the need for clear regulatory frameworks to govern AI deployment. In Europe, for example, the European Commission (2021) has introduced guidelines aimed at ensuring the ethical deployment of AI, with a strong emphasis on fostering trust and advancing human-centric

AI development (Vesnic-Alujevic et al. 2020). These frameworks seek to align AI development with fundamental human rights, prioritizing the protection of privacy, ensuring fairness, and enhancing transparency in decision-making processes. The Council of Europe (2018) has further underscored the importance of ensuring that AI systems are designed and operated in a way that upholds these core values, mitigating risks related to bias, inequality, and discrimination.

3.3 Societal Impacts and Public Trust

The transformative potential of AI in reshaping societal functions is accompanied by the risk of deepening existing divides between those who benefit from AI and those who are marginalized by it. One of the most pressing ethical concerns is the erosion of public trust, particularly when AI systems are used to make decisions in sensitive domains like healthcare, where human lives are directly impacted. In such contexts, the transparency of AI systems becomes critical. Concerns about transparency are closely tied to broader issues, such as surveillance, data privacy, and the "black box" nature of many AI systems, which can obscure the underlying logic of their decisions, making it difficult for the public to trust the outcomes (Pflanzer et al. 2023).

Moreover, ethical AI development must prioritize the preservation of human autonomy and dignity, particularly in situations where AI assumes decision-making roles traditionally held by humans. This is especially crucial in morally complex scenarios, such as healthcare decisions or judicial rulings, where human judgment must remain central. AI should be designed to complement, rather than replace, human judgment, ensuring that individuals retain control over decisions that profoundly affect their lives. Preserving human agency in these contexts is not only crucial for maintaining public trust but also for fostering the ethical integration of AI into societal systems.

Figure 3.1.6 outlines the main ethical and policy considerations in AI through four interconnected areas: ethical implications, governance, accountability, and societal impacts and public trust. Ethical implications include ensuring fairness and transparency in AI systems, mitigating bias to prevent discrimination, especially against vulnerable groups, and developing audit trails to support accountability and transparency. Governance focuses on establishing regulatory frameworks, such as European Commission guidelines, to align AI with human rights, while also adapting governance structures to regulate autonomous systems. Accountability concerns the definition of responsibility for AI-driven decisions, including the roles of designers and data providers. Societal impacts and public trust address concerns related to data privacy, "black box" systems, human autonomy, dignity, and the need for AI to complement human judgment in morally complex domains such as healthcare and judicial decision-making. Overall, the figure presents the ethical and policy landscape of AI as a framework linking fairness, governance, responsibility, public trust, and human-centered decision-making.

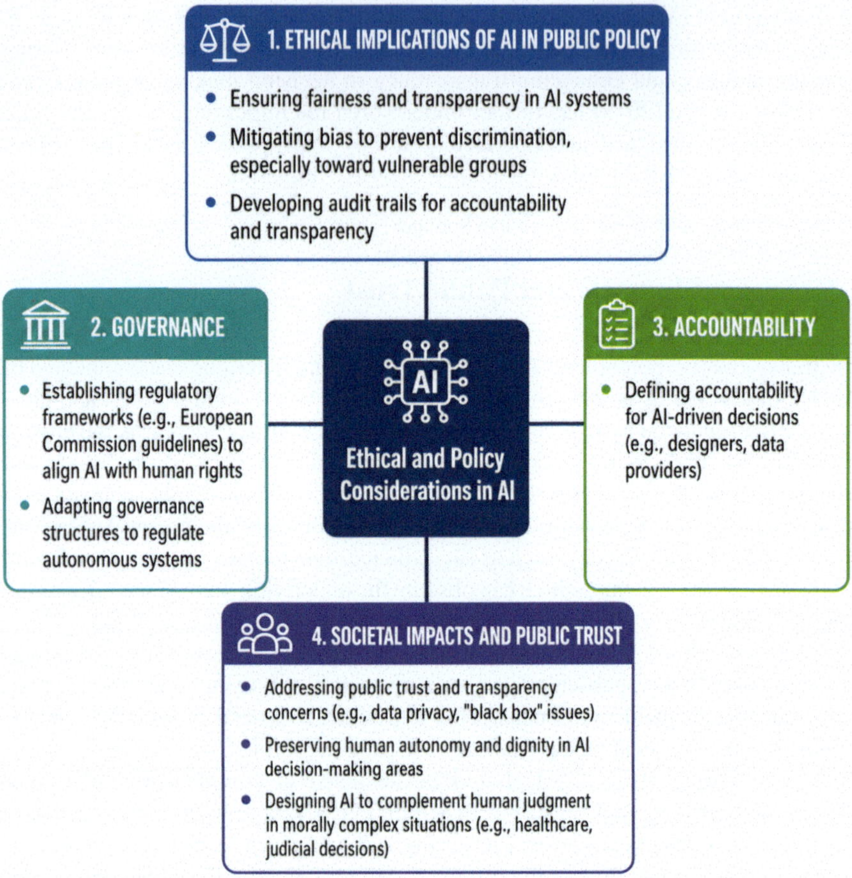

Fig. 3.1.6 Ethical and policy considerations in AI

3.4 Ethical Challenges in AI and Privacy Concerns

In the context of AI-driven public services, ethical challenges concerning privacy, particularly data collection and surveillance in urban environments, have emerged as critical issues. The integration of AI technologies into urban systems, such as traffic management, energy consumption, and waste management, has significantly increased data collection from citizens, raising concerns about potential surveillance overreach and the misuse of personal information (Belli et al. 2020). In smart cities, AI-enabled systems rely on data from a network of sensors and devices, often without the direct consent or awareness of individuals. This results in pervasive data gathering that can infringe on privacy rights by continuously monitoring citizens' behaviors, movements, and interactions (Williamson and Prybutok 2024). To safeguard individual privacy within AI-powered urban infrastructures, we propose a formal privacy-preserving framework based on differential privacy (DP). Differential privacy provides a mathematically rigorous guarantee that the output of an AI system remains nearly indistinguishable regardless of whether any single individual's data is included in the training process. This is particularly relevant for smart city applications where continuous data collection from sensors, mobile devices, and embedded systems creates persistent risk of personal exposure.

At the core of this model is the differential privacy guarantee expressed by:

$$\forall S \subseteq O, Pr[M(D_1) \in S] \leq e^{\epsilon} \cdot Pr[M(D_2) \in S]$$

where: $\mathcal{M}$ is the randomized AI mechanism (e.g., a trained model), D_1 and D_2 are two neighboring datasets that differ by only one individual's data, $\mathcal{O}$ is the output space of the mechanism, and ϵ is the privacy budget, which controls the maximum allowable difference in output distributions. This equation ensures that the presence or absence of any individual record in the dataset does not significantly influence the outcome of the AI model, thus preserving user anonymity. The smaller the value of ϵ, the stronger the privacy guarantee, though this typically comes at the cost of some model utility.

To apply this in smart cities, differential privacy can be implemented during AI model training using noise injection. For example, when computing gradients during stochastic optimization (e.g., in neural networks), noise calibrated to the sensitivity of the data is added:

$$\tilde{\nabla} L = \nabla L + N(0, \sigma^2 I)$$

where: $\nabla\mathcal{L}$ is the true gradient of the loss function, $\tilde{\nabla}\mathcal{L}$ is the privacy-preserving noisy gradient used for training, and $\mathcal{N}(0, \sigma^2 I)$ is zero-mean Gaussian noise with covariance $\sigma^2 I$ added to obscure individual contributions. This approach allows AI systems to learn useful urban policies (e.g., traffic flow optimization, energy distribution, or waste scheduling) without compromising individual privacy ensuring that decisions are made collectively, not at the cost of personal surveillance.

Differential privacy enables public AI systems to operate transparently and accountably without compromising the autonomy and rights of individuals. By mathematically bounding the influence of any single citizen's data, it provides a defensible solution to the surveillance risks inherent in smart city applications. When embedded in AI models used for real-time urban management, DP transforms raw sensor and behavioral data into statistically safe inputs ensuring that privacy is not sacrificed for optimization. This model bridges the gap between algorithmic intelligence and civil liberty, offering a principled path for ethical AI deployment in the data-intensive environments of modern cities.

Framework: Privacy-Preserving AI Framework for Urban Systems (PPAI-US)
Layered AI Architecture for Differential Privacy in Smart City Infrastructure

Layer	Main functionality	AI module(s) & components	Operational scope (input → processing → output)
Layer 1	Privacy Risk Identification In Urban AI data collection	Privacy risk assessment engine	Urban sensor data collection maps → Risk profiling of surveillance intensity → Privacy-critical input flagged
Layer 2	Privacy guarantee via ε-differential privacy mechanism	ε-Differential privacy mechanism	Original dataset and privacy budget → Randomized mechanism with DP guarantees → Protected data exposure distribution
Layer 3	Model training using DP-SGD with noise injection	DP-SGD (differentially private stochastic gradient descent)	Model training gradients and sensitivity metrics → Gaussian noise calibration and injection → Privately trained urban AI model
Layer 4	Transparency reporting and citizen oversight mechanism	Explainable privacy audit dashboard	Privately trained model and DP parameters → Interpretability dashboard → Privacy assurance and public transparency reports

The concept of privacy in these systems extends beyond individual data rights to include broader societal implications. As cities become increasingly interconnected, the aggregation of data can lead to surveillance practices that may disproportionately affect marginalized groups, thus exacerbating social inequalities. These concerns underscore the necessity for robust regulatory frameworks that address the unique privacy risks posed by AI-driven public services (King and Meinhardt 2024). To ensure rigorous privacy protections in data-intensive urban environments, the foundation of our framework is grounded in ε-differential privacy. This mechanism mathematically guarantees that no single individual's data has a significant influence on the output of an AI system, thus safeguarding personal anonymity even in the presence of pervasive data collection.

We begin with the standard formalization of the differential privacy constraint:

$$\forall S \subseteq O, Pr[M(D_1) \in S] \leq e^{\varepsilon} \cdot Pr[M(D_2) \in S]$$

where: S is any measurable subset of possible outputs, O is the output space, M is the randomized AI mechanism, D_1 and D_2 are neighboring datasets differing by a single record, ε is the privacy budget parameter. This function bounds the likelihood of any output changing due to a single individual's inclusion, ensuring privacy consistency across the dataset.

While this ε-DP guarantee governs the statistical indistinguishability of outputs, its enforcement during AI model training is operationalized through differentially private stochastic gradient descent (DP-SGD). This method injects calibrated Gaussian noise directly into the training gradients, thereby propagating privacy into the learning process while preserving model convergence.

The noise-injected gradient update rule is defined as follows:

$$\tilde{\nabla} L = \nabla L + N(0, \sigma^2 I)$$

where: ∇L is the true gradient of the loss function, $\tilde{\nabla} L$ is the privacy-preserving noisy gradient, $\mathcal{N}(0, \sigma^2 I)$ is zero-mean Gaussian noise with covariance matrix $\sigma^2 I$, σ is the noise scale. This function enforces ε-DP during model updates, enabling urban systems to learn functional policies from citizen-generated data without compromising individual privacy.

However, differential privacy alone does not address structural disparities in how AI surveillance systems might disproportionately affect marginalized populations. To assess and mitigate such impacts, we introduce a Group Fairness Disparity Index, applied to the outputs of surveillance-driven AI systems. This index quantifies whether protected groups (e.g., by ethnicity, income, or geography) experience systematically different rates of monitoring or intervention.

The disparity measure, based on statistical parity difference, is computed as:

$$\Delta_{\text{SPD}} = Pr[\hat{y} = 1 \mid A = 0] - Pr[\hat{y} = 1 \mid A = 1]$$

where: $\hat{y}$ is the binary surveillance decision (e.g., triggered or not), A is the binary protected attribute (e.g., unprivileged $A = 0$, privileged $A = 1$), $\Pr[\hat{y} = 1 \mid A = a]$ denotes the probability of positive surveillance outcome for group $A = a$. This function measures the difference in surveillance activation rates between demographic groups, thus serving as a fairness audit indicator for privacy-preserving systems.

By integrating differential privacy at the statistical level, DP-SGD at the training level, and group fairness auditing at the decision output level, this architecture ensures not only personal anonymity but also structural equity in how smart urban systems operate. The mathematical flow reflects a principled transition from privacy assurance to fairness verification fortifying public trust in AI-powered civic infrastructures.

Framework: Privacy-Governance AI Framework for Urban Systems (PGAI-US)
Layered AI Architecture for Ethical Surveillance, Privacy, and Regulatory Alignment in Smart Cities

Layer	Main Functionality	AI Module(s) & Components	Operational Scope (Input → Processing → Output)
Layer 1	Privacy risk identification in urban AI data collection	Privacy risk assessment engine	Urban sensor data collection maps → Risk profiling of surveillance intensity → Privacy-critical input flagged

(continued)

(continued)

Layer	Main Functionality	AI Module(s) & Components	Operational Scope (Input → Processing → Output)
Layer 2	Privacy guarantee via ε-differential privacy mechanism	ε-differential privacy mechanism	Original dataset and privacy budget → Randomized mechanism with DP guarantees → Protected data exposure distribution
Layer 3	Model training using DP-SGD with noise injection	DP-SGD (differentially private stochastic gradient descent)	Model training gradients and sensitivity metrics → Gaussian noise calibration and injection → Privately trained urban AI model
Layer 4	Bias-aware surveillance disparity detection	Surveillance disparity detection using group fairness metrics	Aggregated surveillance data and population segments → Fairness impact audit → Disparate exposure detection across groups
Layer 5	Policy compliance monitoring and regulatory alignment	Regulatory compliance monitoring module	Urban AI model metadata and legal guidelines → Compliance scoring and gap identification → Regulatory conformance report
Layer 6	Public transparency, equity auditing, and civic oversight	Civic transparency and oversight platform	Audit results and civic stakeholder input → Equity and governance dashboard → Policy-refined deployment feedback

Existing privacy laws, such as the General Data Protection Regulation (GDPR) in the European Union, provide some protections by ensuring that citizens' data are collected and processed lawfully. The GDPR emphasizes data minimization and the need for explicit consent before data collection can occur, thus safeguarding personal privacy in AI-driven environments (Williamson and Prybutok 2024). However, despite these regulations, there are ongoing debates about whether existing legal frameworks are adequate to address the complexities of AI and the vast scale of data collection that accompanies it. Critics argue that these laws are not fully equipped to handle the systemic risks that AI poses to privacy, particularly in urban settings where data collection is ubiquitous and often passive (King and Meinhardt 2024).

To address the regulatory inadequacies in handling the scale and complexity of AI-driven data collection, we propose a regulation-aware federated learning (RPP-FL) framework. This model allows decentralized AI training across distributed devices such as sensors, smartphones, or municipal databases without transferring raw data to a centralized server. Instead, it exchanges model updates, ensuring compliance with data minimization principles and eliminating the need for passive, non-consensual aggregation. The model explicitly integrates legal constraints into the learning optimization process, operationalizing GDPR principles within the architecture of AI systems.

At the core of this framework lies the constrained federated optimization objective:

$$\min_{\{\theta_i\}} \sum_{i=1}^{N} w_i \mathcal{L}_i(\theta_i) \text{ subject to } D_i^{access} \subseteq D_i^{consent}$$

where: θ_i are the local model parameters for client i, $\mathcal{L}_i(\theta_i)$ is the local loss function computed on client i's data, w_i is the relative weight of the client in the global update, D_i^{access} is the data actually used in training, and $D_i^{consent}$ is the subset of data for which the user has explicitly granted permission. This constraint enforces that only consented data is used for model training, making the learning process legally compliant. To prevent individual data inference from shared gradients, local differential privacy is applied before transmitting updates:

$$\tilde{\nabla} \mathcal{L}_i = \nabla \mathcal{L}_i + N(0, \sigma^2 I)$$

where: $\nabla \mathcal{L}_i$ is the computed gradient on client i's local data, $\tilde{\nabla} \mathcal{L}_i$ is the noisy gradient shared with the central server, and $\mathcal{N}(0, \sigma^2 I)$ is Gaussian noise injected to preserve individual privacy.

Once updates are collected, a global aggregation function computes the updated central model while preserving fairness and data minimization:

$$\theta^{(t+1)} = \sum_{i=1}^{N} \frac{n_i}{n} \tilde{\nabla} \mathcal{L}_i^{(t)}$$

where: $\theta^{(t+1)}$ is the global model at iteration $t + 1$, n_i is the number of samples at client i, and $n = \sum_{i=1}^{N} n_i$ is the total number of samples across all clients.

This regulation-aware federated learning framework allows public AI systems to comply with legal and ethical data governance principles by enforcing consent-aware training, decentralized computation, and privacy-preserving communication. It directly addresses the core limitations of current privacy laws in large-scale urban AI deployments by operationalizing GDPR principles such as data minimization and informed consent within the very structure of AI training. In doing so, it ensures that public AI systems are not only efficient and scalable but also legally compliant and socially accountable, bridging the divide between regulatory intention and algorithmic implementation.

Framework: Regulation-Aware Privacy-Preserving Federated Learning Framework (RPP-FL)

Layered AI Architecture for GDPR-Compliant Federated Intelligence

Layer	Main functionality	AI module(s) & components	Operational scope
Layer 1	Consent-aware data input and governance initialization	Consent validation and governance engine	User-consented data → Validation of legal constraints → Legally compliant training dataset
Layer 2	Federated model training with regulatory constraints	Federated constrained optimization	Legal-compliant local data → Federated training with consent constraint → Local model updates

(continued)

(continued)

Layer	Main functionality	AI module(s) & components	Operational scope
Layer 3	Local privacy-preserving gradient obfuscation	Local differential privacy (LDP) gradient perturbation	Local gradients → Noise injection via LDP → Obfuscated gradients
Layer 4	Federated aggregation with fairness constraints	Privacy-aware federated aggregation	Obfuscated gradients from clients → Federated aggregation → Updated global model
Layer 5	Post-training governance and legal compliance reporting	Legal compliance audit module	Updated global model → Evaluation against legal standards → Governance-compliant audit output

Moreover, the effectiveness of privacy laws in protecting citizens relies heavily on the transparency and accountability of AI systems. Without clear oversight and mechanisms to ensure that AI technologies respect privacy rights, the potential for misuse remains significant. Additionally, the concept of privacy by design, which integrates privacy protections into the development of AI systems from the outset, is seen as a crucial strategy for mitigating privacy risks in smart cities (Belli et al. 2020). To operationalize the principle of Privacy by Design in AI systems used in smart cities, we introduce a formal privacy-aware architecture built on differential privacy (DP). Unlike post-deployment retrofits, this model embeds privacy constraints directly into the AI development lifecycle from data collection and processing to training, inference, and decision output. It ensures that privacy protections are not reactive measures, but proactively engineered into every computational component of the system.

At the foundation of this model is the standard differential privacy constraint, which applies globally to any data processing mechanism $\mathcal{M}$:

$$\forall S \subseteq O, Pr[M(D_1) \in S] \leq e^{\epsilon} \cdot Pr[M(D_2) \in S]$$

where: $\mathcal{M}$ is any AI system component accessing the data, D_1 and D_2 are neighboring datasets differing by one record, $\mathcal{O}$ is the space of outputs, and $\epsilon\epsilon$ is the privacy budget that bounds the divergence in system outputs due to any individual's data.

To enforce this constraint during model training in smart city systems, a privacy-aware gradient mechanism is implemented:

$$\tilde{\nabla} L = \text{Clip}(\nabla L, C) + N(0, \sigma^2 I)$$

where: $\nabla\mathcal{L}$ is the raw gradient, Clip $(\cdot, C)$ limits the gradient norm to threshold C to control sensitivity, and $\mathcal{N}(0, \sigma^2 I)$ is Gaussian noise added to obscure individual influence.

Finally, the privacy guarantees are tracked and managed through a cumulative privacy accountant, which ensures that the overall system remains within its allowable privacy budget across all operations:

$$\epsilon_{total} = \sum_{t=1}^{T} \epsilon_t \leq \epsilon_{max}$$

where: ϵ_t is the privacy cost at time t, T is the total number of AI operations, and ϵ_{max} is the maximum allowable cumulative privacy loss for the system.

This privacy-by-design framework embeds differential privacy guarantees at the architectural level, ensuring that AI systems used in smart cities are not only effective but legally and ethically accountable. By formally controlling the influence of each data subject throughout training and inference, the model preempts potential misuse and builds institutional trust in AI governance. It allows public agencies to audit privacy costs, certify compliance with regulatory frameworks such as GDPR, and implement AI that is transparent, explainable, and socially responsible by design.

Framework: Differential Privacy-By-Design Smart City AI Framework (DPDSC-AIF)
Layered AI Architecture for Legally Accountable Privacy-Embedded Intelligence

Layer	Main functionality	AI module(s) & components	Operational SCOPE
Layer 1	Legal compliance initialization and privacy risk contextualization	Legal privacy context engine	Privacy legislation, system specifications → Contextual privacy framework → Design inputs
Layer 2	Differential privacy constraint enforcement in AI mechanisms	ε-Differential privacy constraint	Data access modules → Constraint enforcement → Privacy-bounded operations
Layer 3	Gradient sensitivity control with privacy-preserving noise	DP-stochastic gradient descent (DP-SGD)	Model gradients → Clipping and noise injection → Obfuscated learning updates
Layer 4	Cumulative privacy budget management and tracking	Privacy accountant	Per-operation privacy cost → Budget tracking → Auditable system-wide privacy loss
Layer 5	Post-training certification and governance compliance	Governance compliance certification module	AI system outputs → Legal verification → Privacy certification reports

While AI-driven public services offer significant benefits in optimizing resource management and improving efficiency, they also introduce substantial ethical challenges related to privacy. Stronger regulatory measures, transparent governance, and the adoption of privacy-preserving technologies are essential to ensuring that the integration of AI into public services does not come at the cost of individual and societal privacy rights.

3.5 Conclusion

This chapter demonstrates that the ethical and policy challenges of artificial intelligence cannot be treated as peripheral safeguards; they must be designed into the very architecture of AI systems that govern public life. It has shown that fairness, transparency, accountability, and privacy are not aspirational ideals but enforceable constraints that can be operationalized through rigorous mathematical formulations, regulatory frameworks, and governance mechanisms. By embedding these values at every stage of the AI lifecycle from data collection and training to decision-making and oversight, societies can prevent technology from amplifying existing inequities and instead channel its power toward inclusion and justice.

The chapter makes clear that privacy-preserving methods such as differential privacy, federated learning, and privacy-by-design architectures are not optional add-ons but essential infrastructures for safeguarding individual rights in data-intensive environments. Likewise, fairness-constrained optimization, disparity audits, and accountability dashboards provide governments and institutions with the means to detect, measure, and correct bias before it corrodes public trust. At the same time, human judgment must remain sovereign: AI should support professional discretion, never displace it, particularly in morally sensitive domains like healthcare and justice.

The analysis reveals that the future of AI in governance will be defined not by how fast systems can compute but by how faithfully they embody democratic values. Without privacy, accountability, and equity engineered into their design, AI systems risk encoding discrimination, exclusion, and surveillance into the fabric of society. But when these principles are mathematically grounded and institutionally enforced, AI becomes more than a tool of efficiency: it becomes a civic utility—transparent, equitable, and trustworthy. The chapter thus offers both a warning and a blueprint, urging policymakers, researchers, and technologists to ensure that the digital infrastructures of tomorrow are worthy of the democratic societies they serve.

References

Belli, L., Cilfone, A., Davoli, L., Ferrari, G., Adorni, P., Di Nocera, F., Dall'Olio, A., Pellegrini, C., Mordacci, M., & Bertolotti, E. (2020). IoT-enabled smart sustainable cities: Challenges and approaches. *Smart Cities*. https://doi.org/10.3390/smartcities3030052

European Commission, (2021). *AI Watch: Artificial Intelligence in public services*. Publications Office of the European Union. https://ai-watch.ec.europa.eu/news/beyond-pilots-sustainable-implementation-ai-public-services-ai-watch-report-2021-10-27_en

Council of Europe, (2018). Algorithms and human rights - Study on the human rights dimensions of automated data processing techniques and possible regulatory implications.

King & Meinhardt (2024). Rethinking privacy in the AI era: Policy provocations for a data-centric world. *Stanford Institute for Human-Centered Artificial Intelligence*. https://hai.stanford.edu/assets/files/2024-02/White-Paper-Rethinking-Privacy-AI-Era.pdf

Pflanzer, M., Dubljević, V., Bauer, W.A. et al. Embedding AI in society: ethics, policy, governance, and impacts. AI & Soc 38, 1267–1271 (2023). https://doi.org/10.1007/s00146-023-01704-2

Sargiotis, D. (2024). Ethical AI in Information Technology: Navigating Bias, Privacy, Transparency, and Accountability. *Advances in Machine Learning & Artificial Intelligence*, 5(3), 1–14. https://doi.org/10.33140/AMLAI.05.03.03

Vesnic-Alujevic, L., Nascimento, S., & Pólvora, A. (2020). Societal and ethical impacts of artificial intelligence: Critical notes on European policy frameworks. *Telecommunications Policy, 44*(4), 101961. https://doi.org/10.1016/j.telpol.2020.101961

Williamson, S. M., & Prybutok, V. (2024). Balancing privacy and progress: A review of privacy challenges, systemic oversight, and patient perceptions in AI-driven healthcare. *Applied Sciences.* https://doi.org/10.3390/app14020675

Chapter 4
Mitigating Data Breach Risks in AI Systems for Public Services: Best Practices and Strategies

This chapter confronts a hard truth: in public services, an AI system is only as trustworthy as its defenses. It maps the full risk surface, data poisoning during training, adversarial perturbations at inference, and privacy-eroding inference attacks and pairs each with deployable, mathematically grounded countermeasures for real-world ministries, cities, and agencies.

First, it addresses poisoned learning pipelines by casting training as a minimax game against worst-case corrupted distributions and by using influence-function screening to detect and down-weight malicious samples so eligibility, triage, and allocation engines remain reliable even when a fraction of the data is compromised.

Second, it hardens models against adversarial examples through ℓp-bounded adversarial training, projected-gradient adversaries, and first-order certification that quantifies prediction stability an essential guardrail for health diagnostics, social benefits, and emergency dispatch where subtle input nudges can misroute scarce resources.

Third, it tackles inference-time privacy leakage by enforcing output-layer differential privacy and calibrated noise mechanisms, limiting what repeated queries can reveal about any individual thus preserving confidentiality without sacrificing service utility.

Across these pillars, the chapter packages practice-ready frameworks ARPAI-DF for poisoning resilience, CARPAF for certified robustness, and ISPAIPF for output privacy, each coupled to governance controls: fairness and robustness audits, privacy accountants, and compliance dashboards that institutionalize accountability. The result is a blueprint for resilient, rights-preserving AI in public infrastructure: fast enough for real-time governance, rigorous enough for academic scrutiny, and transparent enough to earn public trust.

D. Sargiotis, *Empathy in Algorithms*, https://doi.org/10.1007/978-3-032-32128-2_4

4.1 Data Poisoning

One of the most significant risks in AI systems is data poisoning, where malicious actors manipulate the training data used in machine learning models. This can lead to the AI making inaccurate predictions or decisions, particularly in critical public sectors such as healthcare or social welfare. Data poisoning can result in improper service allocation or biased outcomes, compromising the integrity of public services (Villegas-Ch and García-Ortiz 2023). To preserve the reliability and fairness of AI systems in public domains, we propose a robust adversarial training framework that defends against data poisoning attacks. This model accounts for the possibility that a fraction of the training data may be corrupted by adversaries seeking to manipulate the AI's predictions. By formalizing training as a minimax optimization problem, the system minimizes its loss under the worst-case perturbation of the data distribution, ensuring resilience even in hostile environments.

The core formulation of this framework is defined as:

$$\underset{\theta}{min}\underset{\delta\in\Delta}{max}\mathbb{E}_{(x,y)\sim D+\delta}[L(f_\theta(x), y)]$$

where: θ represents the model parameters, $\mathcal{L}$ is the loss function (e.g., cross-entropy), D is the clean data distribution, δ is an adversarial perturbation in the poisoning set Δ, and $f_\theta(x)$ is the AI model's prediction given input xx.

This formulation ensures that the model optimizes performance against the worst-case poisoned distribution $D+\delta$, forcing it to learn representations that are less sensitive to anomalous or malicious samples. The constraint set ΔΔ defines permissible manipulations based on either a bounded proportion of poisoned data or perturbation norms.

To improve computational efficiency, a surrogate formulation using influence functions can approximate the effect of poisoned points on the model's prediction:

$$I_{(x_c,y_c)} = \nabla_\theta L(f_\theta(x_c), y_c)^\top H_\theta^{-1} \nabla_\theta \mathcal{L}_{\text{train}}$$

where: $I_{(x_c,y_c)}$ estimates the influence of a candidate training point (x_c, y_c) on the total training loss, H_θ is the Hessian matrix of second-order derivatives with respect to θ, and $\nabla_\theta \mathcal{L}_{\text{train}}$ is the gradient over clean training data. High-magnitude influence values can signal potential poisoning points, which are then excluded or reweighted to minimize their corruptive impact.

This adversarially training framework defends AI systems against manipulated training data by explicitly modeling the worst-case behavior of poisoned distributions. It ensures that public service algorithms maintain reliable performance even when a subset of data has been tampered with. In sensitive domains such as healthcare or social welfare, this defense mechanism safeguards the integrity of service allocation, protects vulnerable populations from algorithmic manipulation, and reinforces institutional trust in AI governance. By anticipating and resisting adversarial interference, the model transforms AI into a resilient component of secure and equitable public infrastructure.

Framework: Adversarially Robust Public AI Defense Framework (ARPAI-DF)
Layered AI Architecture for Poison-Resilient Decision Systems in Public Services

Layer	Main functionality	AI module(s) & components	Operational scope
Layer 1	Threat context initialization and poisoning risk framing	Adversarial threat modeling engine	Public domain AI context → Identification of poisoning risk → Robust design objectives
Layer 2	Robust model training against poisoned distributions	Adversarial training for data poisoning robustness	Training data (clean + adversarial) → Minimax optimization → Poison-resilient model weights
Layer 3	Influence function-based poison detection and reweighting	Influence function-based poison detection	Training sample gradients → Influence estimation → Outlier detection and data reweighting
Layer 4	Model integrity validation and governance reporting	Robustness compliance audit module	Final model predictions → Bias inspection and robustness verification → Compliance reporting

4.2 Adversarial Attacks

Another considerable risk is adversarial attacks, where attackers introduce slight alterations in the input data, which can mislead AI models into making incorrect predictions. In the context of public services, adversarial attacks can severely impact resource allocation, creating unfair or inefficient outcomes (Sangwan et al. 2023). To mitigate the threat of adversarial attacks in AI-driven public service systems, we propose a certified robustness framework based on adversarial training with ℓp-bounded perturbations. This model ensures that predictions remain stable under small, adversarially crafted perturbations in the input data. In contexts such as public health triage or emergency resource dispatch, this robustness is essential to maintain fair and efficient allocation of limited resources even when input signals may be subtly manipulated by malicious actors.

The robust training objective is formalized as a minimax optimization problem:

$$\underset{\theta}{min}\mathbb{E}_{(x,y)\sim D}[\underset{\|\delta\|_p\leq\epsilon}{max}\,L(f_\theta(x+\delta),y)]$$

where: θ are the model parameters, D is the clean training distribution, $f_\theta(x)$ is the model's prediction for input x, δ is an adversarial perturbation bounded in norm

$\| \cdot \|_p$ by radius ϵ, and $\mathcal{L}$ is the loss function (e.g., cross-entropy). This objective enforces that the model performs well not only on clean inputs but also on the worst-case perturbation δ within a constrained norm ball, thereby ensuring adversarial robustness.

To enhance provability of robustness, a certified bound on the margin of prediction stability can be derived using first-order Taylor approximation:

$$f_\theta(x+\delta) \approx f_\theta(x) + \nabla_x f_\theta(x) \cdot \delta$$

where: $\nabla_x f_\theta(x)$ is the input gradient of the model. This formulation can be used to estimate how sensitive the model is to adversarial shifts in the input and to identify regions where certification of prediction stability is possible.

Additionally, defensive training includes projected gradient descent (PGD) adversaries, which iteratively maximize the inner loss function to simulate worst-case attacks during training.

This adversarially trained AI model ensures that public service systems remain stable and accurate even when input data is subtly manipulated by adversaries. By learning representations that are inherently robust to small perturbations, the model guarantees reliable performance in security-sensitive contexts such as healthcare diagnostics, social benefit eligibility, or traffic management. It transforms AI from a vulnerable prediction engine into a resilient infrastructure tool, ensuring that adversarial tampering cannot undermine fairness, efficiency, or trust in automated decision-making. This certified robustness framework is foundational for deploying AI securely in open, adversarial urban environments.

Framework: Certified Adversarial Robustness Public AI Framework (CARPAF)
Layered AI Architecture for Secure and Fair AI in Public Decision Systems

Layer	Main functionality	AI module(s) & components	Operational scope
Layer 1	Threat modeling and service vulnerability assessment	Adversarial vulnerability analysis engine	Public service domains → Identification of adversarial attack risks → Robustness requirement specification
Layer 2	Robust model training with ℓp-bounded perturbations	ℓp-bounded adversarial training	Clean training data → Adversarial loss minimax training → Robust prediction parameters
Layer 3	Prediction stability certification via input gradient approximation	First-order input gradient certification	Model predictions → Taylor-based gradient evaluation → Certified robustness bounds
Layer 4	Adversarial example simulation via gradient-based perturbation	Projected gradient descent adversarial generation	Input samples → Iterative adversarial perturbation → Synthetic attack generation for training

(continued)

(continued)

Layer	Main functionality	AI module(s) & components	Operational scope
Layer 5	Post-training fairness and robustness audit	Fairness and robustness compliance module	Model outcomes → Performance and fairness audit → Governance-aligned robustness validation

4.3 Inference Attacks

Inference attacks involve unauthorized access to the AI system's output patterns, enabling attackers to extract sensitive information about the data used to train the model. Public service AI systems that handle personal information (such as healthcare records or financial data) are especially vulnerable to such attacks, which can lead to significant privacy breaches (Villegas-Ch and García-Ortiz 2023).

To defend AI systems in public services from inference attacks, we propose a differential privacy-based inference shielding mechanism. Inference attacks exploit the statistical patterns in model outputs to reconstruct or infer properties of individual training records. This is particularly dangerous in public domains like healthcare or finance, where AI outputs even when anonymized can unintentionally leak sensitive attributes. By applying differential privacy to the output layer, we ensure that individual information cannot be inferred from repeated model interactions.

The core guarantee is expressed by the standard $\epsilon\epsilon$-differential privacy condition on the model output $\mathcal{M}$:

$$\forall S \subseteq O, Pr[M(D_1) \in S] \leq e^{\epsilon} \cdot Pr[M(D_2) \in S]$$

where: $\mathcal{M}$ is the AI system's output mechanism (e.g., prediction API), D_1 and D_2 are neighboring datasets differing by one record, $\mathcal{O}$ is the space of possible outputs, and ϵ is the privacy budget that bounds the information leakage between outputs.

To implement this at the inference stage, we introduce randomized response noise into the output vector:

$$\tilde{f}_\theta(x) = f_\theta(x) + N(0, \sigma^2 I)$$

where: $f_\theta(x)$ is the raw model prediction for input x, $\tilde{f}_\theta(x)$ is the privatized output returned to the user or API caller, and $\mathcal{N}(0, \sigma^2 I)$ is zero-mean Gaussian noise added to prevent reconstruction of sensitive information.

This mechanism limits how much any individual output can reveal about the data used to generate it, especially when attackers attempt membership inference or attribute reconstruction via repeated queries.

By embedding differential privacy directly into the AI system's output interface, this model protects against inference attacks without sacrificing functional utility. It ensures that no individual data point disproportionately influences the model's observable behavior, preserving the confidentiality of training data even under adversarial querying. This is critical in public service systems such as digital health assistants or financial eligibility engines where predictions must remain accessible yet secure. The model offers a mathematically provable guarantee of output-level privacy, reinforcing public trust in AI systems that manage sensitive societal functions.

Framework: Inference Shielding Public AI Privacy Framework (ISPAIPF)
Layered AI Architecture for Output Privacy in Public Decision Systems

Layer	Main functionality	AI module(s) & components	Operational scope
Layer 1	Privacy threat modeling and risk definition	Inference threat analysis engine	Public service domains → Identification of inference attack vectors → Privacy safeguarding requirements
Layer 2	Output-level privacy guarantee enforcement	Output-layer differential privacy guarantee	Prediction mechanism → ε-DP constraint application → Controlled information leakage
Layer 3	Output vector obfuscation with noise injection	Gaussian mechanism for output privatization	Model predictions → Gaussian noise addition → Privacy-shielded output
Layer 4	Post-inference privacy compliance and risk auditing	Privacy compliance audit module	Privatized system responses → Output review and leakage analysis → Compliance validation

4.4 Conclusion

This chapter has demonstrated that securing artificial intelligence in public services is not a peripheral technical concern but a fundamental condition for democratic legitimacy and institutional trust. It has shown that without rigorous defenses, AI systems are vulnerable at every stage, susceptible to poisoned data during training, fragile to adversarial perturbations at deployment, and exposed to inference attacks that threaten privacy and transparency. Left unchecked, these weaknesses not only endanger service reliability but also risk amplifying inequity, eroding accountability, and undermining public confidence in government institutions.

By embedding mathematically grounded countermeasures, robust training regimes, adversarial certification, and differential privacy guarantees, the chapter illustrates how vulnerabilities can be transformed into resilience. These protections are not abstract ideals but operational tools that ministries, municipalities, and agencies can deploy to ensure that service pipelines remain accurate, fair, and rights-preserving even under hostile conditions. Frameworks such as ARPAI-DF, CARPAF, and ISPAIPF embody this principle, coupling technical defenses with governance mechanisms like fairness audits, compliance dashboards, and privacy accountants. Together, they demonstrate how robust AI can be institutionalized as both a technological and a civic infrastructure.

The conclusion is unambiguous: trustworthy AI for public services must be designed deliberately, with robustness, fairness, and accountability built into its core architecture. Only then can societies harness the speed and intelligence of AI without surrendering to its risks. In this synthesis, AI ceases to be a vulnerability and becomes a resilient civic utility capable of safeguarding both efficiency and the ethical foundations upon which democratic governance depends.

References

Sangwan, R.S., Badr, Y., & Srinivasan, S.M. (2023). Cybersecurity for AI Systems: A Survey. *Journal of Cybersecurity and Privacy*. https://doi.org/10.3390/jcp3020010

Villegas-Ch, W., & García-Ortiz, J. (2023). Toward a Comprehensive Framework for Ensuring Security and Privacy in Artificial Intelligence. *Electronics*. https://doi.org/10.3390/electronics12183786

Chapter 5
AI Security in Public Services: Encryption, Privacy, and Resilient Governance

This chapter tackles a foundational mandate for modern government: if public services are to be AI-enabled, they must be secure by design. It charts an end-to-end security architecture for data-intensive ministries, cities, and agencies, translating first principles into deployable mechanisms that protect confidentiality, integrity, and public trust. The narrative begins with encryption as a default posture not only in transit and at rest, but during computation so that models can operate on ciphertext without exposing citizen data. It then narrows the human attack surface through role-based, policy-aware access control, binding accountability to permissions at the precise point of use.

From there, the chapter turns to operational hardening. Continuous security auditing replaces periodic checklists with learning systems that surface drift, tampering, and misuse in real time. Model robustness is treated as a requirement, not a luxury: adversarial training and rigorous robustness evaluation stabilize predictions under pressure so that triage, eligibility, and dispatch pipelines do not collapse when inputs are nudged by an adversary. To contain privacy risk in analytics and model training, a dual track of differential privacy and attribute masking enables learning at population scale without re-identification even if datasets are leaked.

Because incidents are inevitable, the chapter formalizes incident response as a decision process under uncertainty, optimizing detection, mitigation, and notification to minimize harm while satisfying legal duties. Finally, where AI intersects with surveillance, it embeds privacy and fairness directly into policy learning, ensuring that safety goals are pursued without drifting into profiling or over-reach. The result is a coherent, standards-ready playbook: confidential by default, resilient in operation, auditable in practice, and worthy of democratic confidence.

D. Sargiotis, *Empathy in Algorithms*, https://doi.org/10.1007/978-3-032-32128-2_5

5.1 Encryption

Implementing encryption mechanisms to protect data both in transit and at rest is essential. Encryption ensures that even if the data is intercepted during transmission, it remains unreadable without the appropriate decryption key (Villegas-Ch and García-Ortiz 2023). To safeguard the confidentiality of citizen data in AI-driven public services, we propose the integration of Homomorphic Encryption (HE) a cryptographic framework that allows computations to be performed directly on encrypted data. Unlike traditional encryption, which requires decryption before processing (thus exposing sensitive content), HE ensures that the data remains secure throughout the entire AI workflow, including during model inference and training. This is particularly valuable for secure data analysis in healthcare, taxation, and law enforcement applications, where confidentiality is paramount and legal compliance (e.g., GDPR, HIPAA) is required.

The core functionality of HE can be defined as:

$$\mathrm{Dec}(f_\theta^{\mathrm{enc}}(\mathrm{Enc}(x))) = f_\theta(x)$$

where $\mathrm{Enc}(x)$ is the encrypted form of input x, f_θ^{enc} is the AI model evaluated over encrypted data, and Dec $(\cdot)$ is the decryption function. This equation guarantees that applying the model to encrypted input, followed by decryption, yields the same result as applying the model to plaintext input preserving functionality while ensuring data privacy.

Homomorphic operations allow for both addition and multiplication on ciphertexts. In partially homomorphic systems, operations may be limited to one type, while Fully Homomorphic Encryption (FHE) supports arbitrary functions. For secure linear predictions in public systems, we may rely on additive HE:

$$\mathrm{Enc}(a + b) = \mathrm{Enc}(a) + \mathrm{Enc}(b)$$

And for secure dot product operations, which are core to AI model predictions:

$$\mathrm{Enc}\left(w^\top x\right) = \sum_{i=1}^{n} w_i \cdot \mathrm{Enc}(x_i)$$

where w_i are the model weights and x_i are the encrypted feature values. This allows encrypted inference without ever revealing the raw input x to the AI model provider.

This homomorphic encryption framework enables AI systems to maintain complete confidentiality of data in transit, at rest, and during computation. By performing encrypted AI operations end-to-end, it prevents sensitive personal data from being exposed at any stage of processing even to the system operator. This is particularly essential for smart cities and digital public services that rely on third-party models but are obligated to protect citizen privacy. The model aligns with both cryptographic rigor and functional utility, making it an ideal privacy-preserving AI architecture for lawful, ethical, and secure deployment in public sector applications.

Framework: Homomorphic Encryption Secure Public AI Framework (HESPAIF)

Layered AI Architecture for Confidential Encrypted Computation in Public Services

Layer	Main functionality	AI module(s) and components	Operational scope
Layer 1	Privacy-first threat modeling and encryption policy setup	Encryption policy orchestration engine	Confidential public datasets → Risk classification and encryption planning → Homomorphic encryption setup
Layer 2	Model evaluation on encrypted input	Homomorphic encrypted inference module	Encrypted input data → Model execution on ciphertexts → Encrypted predictions
Layer 3	Encrypted addition and linear computation	Additive homomorphic operation module	Encrypted data terms → Encrypted addition operations → Secure feature combination
Layer 4	Secure encrypted dot product for prediction	Encrypted linear inference module	Encrypted features and weights → Dot product over ciphertexts → Privacy-preserving prediction results
Layer 5	Post-inference decryption and compliance validation	Secure decryption compliance module	Encrypted outputs → Authorized decryption and results review → Legal compliance and audit reporting

5.2 Access Controls

Limiting access to AI systems through role-based access controls (RBAC) ensures that only authorized personnel can view or manipulate sensitive data. This reduces the risk of insider threats and unauthorized access (Villegas-Ch and García-Ortiz 2023). To secure AI systems in public service environments from unauthorized access and insider misuse, we propose a formal Role-Based Access Control (RBAC) framework enhanced with a policy-aware decision function. This model encodes access permissions as formal mathematical rules evaluated at the point of data access or model execution. It ensures that only users with appropriate roles can view, modify, or operate on sensitive data and functions safeguarding critical infrastructures such as healthcare records, taxation systems, or smart city control interfaces.

The core access decision function is defined as:

$$A(u, r, o) = \begin{cases} 1, \text{ if } (r, o) \in P_u \\ 0, \quad \text{otherwise} \end{cases}$$

where $A(u, r, o)$ is the access control decision for user u requesting operation oo under role r, and P_u is the set of permissions assigned to u's authenticated role. The function returns 1 (grant) if the role-operation pair (r, o) is authorized under the policy P_u,, and 0 (deny) otherwise.

To formalize access policies system-wide, we define the access rule base as a tuple:

$$P_u = \{(r_i, o_j) | r_i \in R_u, o_j \in O, (r_i, o_j) \in \Pi\}$$

where R_u is the set of roles assigned to user u, O is the set of all operations in the system, and Π is the global access policy that maps roles to allowed operations.

To enforce real-time constraints and auditability, the access control function can be extended with time-based or contextual policies using an evaluation function:

$$A\prime(u, r, o, t, c) = A(u, r, o) \cdot \phi(t, c)$$

where t is the time of request, c is the context (e.g., device ID, location, security level), and $\phi(t, c)$ is a Boolean function that returns 1 only if time and context satisfy predefined constraints.

This policy-constrained access control framework enables secure, role-sensitive operation of AI systems in complex public service environments. By embedding formal authorization logic into the computational fabric of AI, the model prevents misuse by unauthorized actors, enforces institutional boundaries, and supports auditability and compliance. When deployed in public health, welfare, or municipal infrastructures, this model ensures that AI functions remain transparent, traceable, and restricted to legitimate actors safeguarding citizen data and institutional integrity in digital governance systems.

Framework: Role-Based Access Control Policy-Enforced AI Framework (RBAC-PEAF)

Layered AI Architecture for Access-Constrained AI Governance in Public Systems

Layer	Main functionality	AI module(s) and components	Operational scope
Layer 1	Insider threat modeling and role taxonomy initialization	Role taxonomy threat analysis engine	Public service environment → Threat scenarios and user-role classification → Role-policy mapping initialization
Layer 2	Formal access decision enforcement	Role-based access control decision function	User role and operation request → Authorization check → Grant or deny decision

(continued)

(continued)

Layer	Main functionality	AI module(s) and components	Operational scope
Layer 3	Access policy rule base construction	Role-operation permission mapping engine	Global policy set → Role-operation tuple construction → Enforced access policy
Layer 4	Contextual access enforcement and auditability	Context-aware policy evaluation function	User access request + time/context → Conditional validation → Logged and constrained access
Layer 5	Post-access compliance and governance logging	Access governance audit module	Access decisions and usage events → Log storage and audit trail → Transparency and regulatory validation

5.3 Regular Security Audits

Conducting routine security audits helps identify vulnerabilities within AI systems. These audits should involve reviewing the AI model's behavior, analyzing input data for signs of tampering, and testing the system for weaknesses that could be exploited by attackers (Sangwan et al. 2023). To support routine and automated security audits in AI-driven public infrastructures, we propose an anomaly detection framework based on unsupervised statistical modeling of AI input–output behavior. This approach learns normal system behavior during a safe baseline phase, then continuously monitors for deviations in input distributions, gradient flows, or model outputs that may indicate data tampering or adversarial manipulation. It enables real-time auditing by flagging anomalies that fall outside of learned operational norms, allowing security teams to preemptively identify vulnerabilities or ongoing attacks.

To learn $p(x)$, we fit a density model such as a Gaussian Mixture Model (GMM) to the input or gradient space:

$$p(x) = \sum_{k=1}^{K} \pi_k \cdot \mathcal{N}(x|\mu_k, \Sigma_k)$$

where $\mathcal{N}(x|\mu_k, \Sigma_k)$ is the Gaussian density for component k, μ_k and Σ_k are the mean and covariance, and π_k is the mixing weight for each component. This model approximates the distribution of safe inputs or system states under non-adversarial conditions.

At the core of this framework lies a probabilistic anomaly score function:

$$s(x) = -logp(x)$$

where x is the observed input or model behavior pattern, $p(x)$ is the probability density of x under a learned distribution of normal behavior, and $s(x)$ is the anomaly score. Higher values of $s(x)$ indicate low-likelihood events and potential security risks.

At runtime, each incoming input is evaluated using the learned model, and an audit alert is triggered if the anomaly score exceeds a predefined threshold:

$$\text{Alert}(x) = I[s(x) > \tau]$$

where τ is the security threshold for anomaly detection, and $\mathbb{I}[\cdot]$ is the indicator function. If $\text{Alert}(x) = 1$, the input is flagged for security review.

This anomaly detection framework transforms security audits from manual, periodic checks into continuous, real-time AI monitoring processes. By learning a probabilistic representation of safe system behavior and identifying deviations through anomaly scores, the model enables institutions to proactively detect tampering, adversarial attacks, or system failures. It supports public sector AI deployments in healthcare, finance, and infrastructure by ensuring that operational integrity is preserved and that security vulnerabilities are rapidly surfaced for intervention upholding trust and accountability in AI governance systems.

Framework: Probabilistic Anomaly Audit Framework for Public AI Security (PAAFPS)

Layered AI Architecture for Real-Time Security Auditing in Public Service Systems

Layer	Main functionality	AI module(s) and components	Operational scope
Layer 1	Security threat profiling and baseline data collection	Security baseline profiling engine	Initial system inputs and behavior logs → Safe baseline capture → Data for training
Layer 2	Unsupervised behavior modeling and density estimation	Gaussian mixture density modeling module	Baseline data → Probabilistic model fitting → Learned normal behavior distribution
Layer 3	Real-time statistical anomaly scoring	Statistical anomaly scoring module	Incoming input or behavior → Likelihood evaluation → Anomaly score output
Layer 4	Threshold-based security alerting and signal generation	Real-time anomaly alerting module	Anomaly score → Threshold comparison → Triggered alert if abnormal
Layer 5	Post-audit governance review and intervention logging	Governance forensic audit module	Security events and alerts → Review and forensic analysis → Logged interventions and governance updates

5.4 Adversarial Training

Training AI models to recognize and defend against adversarial inputs can improve their resilience. By simulating potential attacks during the training phase, AI models become better equipped to handle real-world adversarial attempts (Sangwan et al. 2023). To improve the robustness of AI systems against adversarial perturbations, we implement Adversarial Training via Minimax Optimization a defense mechanism where AI models are trained not only on clean data but also on synthetic adversarial examples crafted during training. This approach simulates real-world attacks by maximizing the model's loss within a constrained perturbation set and subsequently training the model to minimize that worst-case loss. It is one of the most effective and mathematically sound strategies to prepare AI systems for deployment in security-sensitive public domains.

The training objective is formulated as a minimax problem:

$$\underset{\theta}{min}\mathbb{E}_{(x,y)\sim D}[\underset{\|\delta\|_p\leq\epsilon}{max}\, L(f_\theta(x+\delta), y)]$$

where θ are the model parameters, (x, y) are training samples from distribution D, δ is the adversarial perturbation bounded by norm $\|\delta\|_p \leq \epsilon$, and $\mathcal{L}\ (\cdot)$ is the loss function (e.g., cross-entropy). The inner maximization identifies the most damaging perturbation within an allowed perturbation radius $\epsilon\epsilon$, while the outer minimization adjusts the model to be resilient against such worst-case attacks.

During training, adversarial examples are generated using iterative methods such as Projected Gradient Descent (PGD):

$$\delta^{(t+1)} = \Pi_{\|\delta\|_p\leq\epsilon}[\delta^{(t)} + \alpha \cdot \text{sign}(\nabla_x L(f_\theta(x+\delta^{(t)}), y))]$$

where $\delta^{(t)}$ is the perturbation at iteration t, α is the step size, $\Pi_{\|\delta\|_p\leq\epsilon}$ projects the perturbation back to the feasible set, and $\nabla_x\mathcal{L}\ (\cdot)$ is the gradient of the loss with respect to the input.

This adversarial training model ensures that AI systems develop resistance to input perturbations by learning to classify both clean and adversarial examples with high accuracy. It prepares the model to withstand real-world threats by repeatedly simulating attacks during training and reinforcing its decision boundaries. In public services such as biometric authentication, fraud detection, or automated welfare allocation, this robustness guarantees that malicious input manipulation cannot compromise fairness, accuracy, or security. The minimax adversarial training framework is a cornerstone of modern AI defense and is essential for trusted and resilient public sector AI deployment.

Framework: Adversarially-Robust AI Security Framework (ARASF)

Layered AI Architecture for Resilient and Secure Public Sector AI Deployment

Layer	Main functionality	AI module(s) and components	Operational scope
Layer 1	Adversarial threat modeling and attack simulation	Adversarial risk simulation engine	Input → Security requirements, threat vectors → Output → Defined attack scenarios for training
Layer 2	Adversarial training using minimax optimization	Adversarial training via minimax optimization	Input → Clean data, simulated attacks → Output → Robust model trained under worst-case perturbations
Layer 3	Iterative generation of perturbations	Projected Gradient Descent (PGD)	Input → Current sample and model → Output → Converged adversarial example for training
Layer 4	Model robustness evaluation and validation	Robustness evaluation audit module	Input → Clean and adversarial test samples → Output → Accuracy, robustness, and threat resistance scores

5.5 Data Anonymization

For public services that handle personal data, anonymizing datasets before using them to train AI models can prevent sensitive information from being exposed during data breaches. Techniques such as differential privacy and data masking are effective in maintaining individual privacy (Villegas-Ch and García-Ortiz 2023). To ensure privacy in AI systems trained on sensitive public sector datasets, we propose a combined anonymization framework that integrates Differential Privacy (DP) and Attribute Masking during preprocessing. This dual-method approach allows public agencies to safely use personal data for AI training while guaranteeing that individual records cannot be reconstructed or reidentified even in the event of a data breach. Differential Privacy introduces statistical uncertainty, while masking removes or obscures high-risk fields prior to data utilization.

To inject uncertainty, noise is added to numerical attributes:

$$\tilde{x} = x + N(0, \sigma^2)$$

where x is a sensitive numerical attribute (e.g., income or age), $\tilde{x}$ is the anonymized version, and $\mathcal{N}(0, \sigma^2)$ is Gaussian noise calibrated to the privacy budget ϵ.

In parallel, categorical fields (e.g., name, ID, ZIP code) are processed through masking:

$$x\prime = \text{mask}(x) = \text{hash}(x) \text{ or } \text{truncate}(x)$$

where $x\prime$ is the masked version of xx, and masking is performed using hashing or generalization (e.g., transforming "1985–07-12" to "1980s").

The core DP guarantee for dataset-level anonymization is expressed as:

$$\forall S \subseteq O, Pr[A(D_1) \in S] \leq e^{\epsilon} \cdot Pr[A(D_2) \in S]$$

where $\mathcal{A}$ is the randomized algorithm (e.g., DP preprocessor), D_1 and D_2 are neighboring datasets differing in one individual, $\mathcal{O}$ is the output space (e.g., preprocessed training data), and ϵ is the privacy budget that bounds individual influence.

This anonymization model ensures that sensitive personal data used in public AI systems remains untraceable, even if training datasets are leaked or attacked. Differential Privacy mathematically guarantees that no single individual meaningfully alters the data analysis, while attribute masking structurally eliminates identifiers before model access. Together, these methods uphold individual privacy in healthcare, taxation, and identity management systems enabling responsible AI innovation in the public sector without compromising data protection principles. This dual-layer anonymization approach is essential for privacy-by-design compliance and societal trust in data-driven governance.

Framework: Privacy-Preserving Public AI Data Framework (P3AIDF)

Layered AI Architecture for Differential Privacy and Attribute Masking in Public Sector AI

Layer	Main functionality	AI module(s) and components	Operational scope
Layer 1	Risk identification and data classification	Data sensitivity classification engine	Input → Public datasets with personal identifiers → Output → Sensitivity-labeled data for anonymization strategy
Layer 2	Numerical attribute anonymization	Gaussian mechanism (differential privacy)	Input → Numerical personal data → Output → Statistically anonymized data with injected uncertainty
Layer 3	Categorical attribute masking	Attribute masking (hashing/ generalization)	Input → Categorical sensitive data → Output → Transformed fields using masking, truncation, or hashing
Layer 4	Post-anonymization privacy guarantee validation	ε-differential privacy guarantee	Input → Preprocessed anonymized dataset → Output → Validated privacy compliance based on DP thresholds
Layer 5	Monitoring and governance of anonymized data use	Anonymized data governance module	Input → AI-trained anonymized datasets → Output → Audits, access logs, and governance compliance reports

5.6 Incident Response Plans

Public service organizations must develop comprehensive incident response plans to react quickly in the event of a data breach. These plans should include protocols for detecting breaches, mitigating the damage, and notifying affected individuals (Villegas-Ch and García-Ortiz 2023). To operationalize and optimize data breach response in AI-driven public service environments, we propose a mathematically formal Incident Response Model based on Markov Decision Processes (MDPs). This model encodes the full lifecycle of a security incident detection, classification, mitigation, and notification as a stochastic process, enabling AI systems to select optimal recovery actions in real time. It supports institutions in reducing breach impact by balancing speed, accuracy, and communication under uncertainty.

An MDP is formally defined by the tuple:

$$M = (S, A, T, R, \gamma)$$

where S is the set of system states (e.g., secure, compromised, mitigated), A is the set of possible actions (e.g., isolate node, patch system, notify users), $T : S \times A \to P(S)$ is the transition probability function, $R : S \times A \to R$ is the reward function (penalizing delay and data loss), $\gamma \in [0,1]$ is the discount factor encoding the urgency of early response.

The optimal incident response policy π^* is obtained by maximizing the expected cumulative reward:

$$\pi^*(s) = \underset{\pi}{argmax} E[\textstyle\sum_{t=0}^{\infty} \gamma^t R(s_t, a_t) | s_0 = s, a_t = \pi(s_t)]$$

where s_t is the system state at time t, $\pi(s_t)$ is the response action selected under policy π, and the expression calculates the expected return under the given policy from initial state s, balancing long-term mitigation effectiveness against immediate impact.

AI systems can be trained on historical breach data and simulated attack scenarios to learn transition probabilities and expected impact. This enables automated response strategies that adapt in real time to the evolving threat landscape and system condition.

This MDP-based framework transforms incident response planning from a static checklist into a dynamic decision-making system. By modeling breach scenarios probabilistically and learning optimal mitigation strategies over time, the model supports rapid, data-driven intervention while accounting for uncertainty, urgency, and resource constraints. Public agencies can deploy this model to ensure that breach detection triggers real-time automated decisions, minimizing damage and accelerating recovery. As part of a comprehensive AI governance strategy, this model ensures resilience, accountability, and compliance with legal obligations to notify affected citizens reinforcing institutional trust in the face of cybersecurity threats.

Framework: AI-Driven Incident Response Optimization Framework (AI-IROF)

Layered AI Architecture for Real-Time Cybersecurity Breach Management in Public Services

Layer	Main functionality	AI module(s) and components	Operational scope
Layer 1	Incident detection and alert generation	Intelligent intrusion detection engine	Input → System logs, intrusion indicators → Output → Triggered alerts and candidate breach events
Layer 2	Incident modeling using stochastic process	Markov Decision Process (MDP)	Input → Security incident scenario data → Output → Encoded states, actions, and transitions for response optimization
Layer 3	Policy optimization for response actions	Optimal policy learning (value iteration/ Q-learning)	Input → MDP state transitions and rewards → Output → Optimal response strategy minimizing breach impact
Layer 4	Real-time response execution and communication	Automated incident response orchestrator	Input → Optimized policy decisions → Output → Isolated systems, notified users, and adaptive mitigation commands

Figure 5.1 presents best practices for ensuring data security in public services through six interconnected areas: encryption, access controls, regular security audits, adversarial training, data anonymization, and incident response plans. Encryption protects data in transit and at rest, ensuring that intercepted data remains unreadable without a decryption key. Access controls, including Role-Based Access Controls (RBAC), limit access to authorized personnel and reduce the risk of insider threats and unauthorized access. Regular security audits identify vulnerabilities within AI systems, review AI model behavior, and test for weaknesses. Adversarial training improves resilience to real-world attacks by training AI systems to recognize and defend against adversarial inputs. Data anonymization protects personal data from exposure through techniques such as differential privacy and data masking. Incident response plans establish protocols for detecting and mitigating breaches, including notifying affected individuals. Overall, the figure presents a layered approach to data security in public services, combining technical protection, controlled access, AI-specific testing, privacy-preserving methods, and breach-response procedures.

The integration of AI systems in public services offers significant potential for enhancing efficiency and service delivery. However, the risks associated with data breaches such as data poisoning, adversarial attacks, and inference threats pose serious challenges to maintaining data security and privacy. By implementing best practices, including encryption, access controls, adversarial training, and regular security audits, public service organizations can mitigate these risks and safeguard the integrity and privacy of their systems.

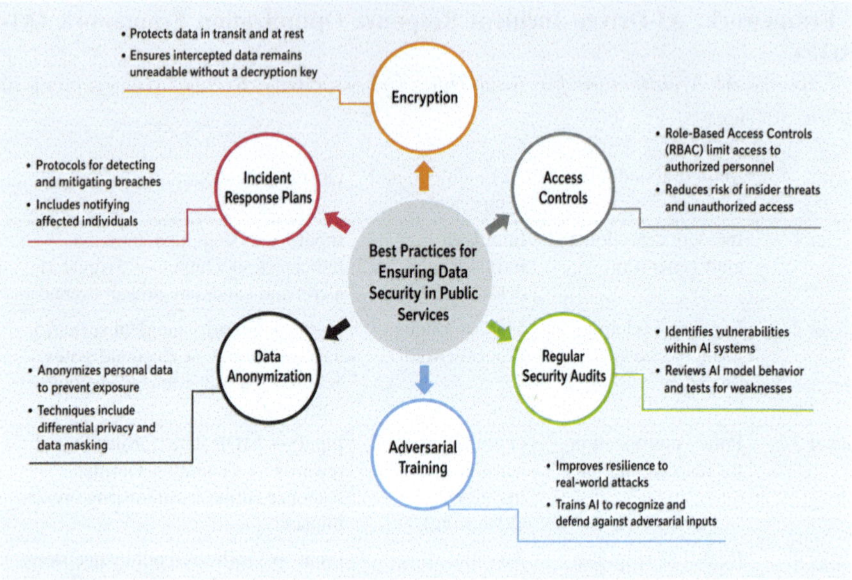

Fig. 5.1 Best practices for ensuring data security in public services

5.7 AI-Driven Surveillance

The increasing reliance on AI-driven surveillance technologies in urban settings raises significant ethical concerns related to privacy, transparency, and the balance between security and individual freedoms. As cities adopt more surveillance systems such as facial recognition and data collection via the Internet of Things (IoT), the ethical question of how these technologies impact personal privacy becomes paramount (Ryan and Gregory 2019). Urban surveillance, often justified by the need for enhanced public safety, risks creating a pervasive environment of constant monitoring. AI systems can potentially identify, track, and profile individuals, which may infringe on citizens' rights to privacy and autonomy (Sanchez et al. 2024). To address the ethical tensions arising from AI-driven surveillance, we propose a Privacy-Preserving Reinforcement Learning Model that embeds differential privacy and adversarial debiasing directly into the policy learning process. This model allows public safety objectives to be pursued while mathematically bounding the risk of personal identification and surveillance-based profiling. Rather than disabling surveillance systems altogether, this framework introduces formally verifiable privacy constraints into decision-making, thus offering a structured path forward that respects individual autonomy while retaining system utility.

The core function of this framework begins with a Differentially Private Policy Learning Model using noise-regularized PPO:

$$\pi_\theta^* = \underset{\theta}{\arg max}\mathbb{E}_t[\frac{\pi_\theta(a_t|s_t)}{\pi_{\theta_{\text{old}}}(a_t|s_t)} \cdot \hat{A}_t - \beta \cdot \mathcal{L}_{\text{privacy}}(s_t, a_t)]$$

where $\pi_\theta(a_t|s_t)$ is the surveillance decision policy, $\hat{A}_t$ is the estimated advantage for selecting action a_t, and $\mathcal{L}_{\text{privacy}}(s_t, a_t)$ is the privacy risk loss associated with that action in state s_t. The scalar β controls the weight of privacy concerns in the learning process.

To explicitly define the privacy risk penalization, we introduce a Differential Privacy Loss Function:

$$\mathcal{L}_{\text{privacy}}(s_t, a_t) = \epsilon^{-1} \cdot \text{KL}(P_{s_t}(a_t) \| P_{s_t}^{\text{noisy}}(a_t))$$

where $P_{s_t}(a_t)$ is the original action probability distribution (e.g., camera activation or facial identification), $P_{s_t}^{\text{noisy}}(a_t)$ is the noise-injected policy output ensuring differential privacy, $\text{KL}(\cdot\|\cdot)$ is the Kullback–Leibler divergence quantifying deviation between original and privatized actions, and ϵ is the differential privacy parameter bounding the probability of identity leakage.

To further ensure fairness in surveillance decisions, particularly in avoiding bias toward marginalized or over-policed populations, we define the Adversarial Debiasing Constraint as:

$$\mathcal{L}_{\text{bias}}(a_t) = \sum_{g \in G} |\mathbb{E}[a_t|g] - \mathbb{E}[a_t]|$$

where g represents a protected demographic group (e.g., ethnicity, disability status), G is the set of all protected groups, and $\mathbb{E}[a_t]$ denotes the expected surveillance action within group g. This bias loss penalizes demographic disparities in surveillance frequency or intensity.

The final Composite Surveillance Ethics-Aware Reward Function becomes:

$$R(s_t, a_t) = U_{\text{safety}}(a_t) - \gamma \cdot \mathcal{L}_{\text{privacy}}(s_t, a_t) - \delta \cdot \mathcal{L}_{\text{bias}}(a_t)$$

where $U_{\text{safety}}(a_t)$ represents the public safety utility of action a_t, γ and δ are regularization parameters for privacy and bias loss respectively.

This AI-driven model creates a structured and ethically grounded decision-making process for urban surveillance systems. By integrating differential privacy through noise injection and bias correction via adversarial constraints, the framework aligns real-time AI actions with both constitutional rights and human-centered ethical design. Rather than treating privacy as a post-hoc filter, this model embeds it directly into the AI's learning dynamics offering a path forward for surveillance that is both intelligent and accountable.

Framework: Privacy-Aligned Surveillance Ethics Reinforcement Framework (PASERF)

Layered AI Architecture for Ethically Governed Privacy-Preserving Urban Surveillance

Layer	Main functionality	AI module(s) and components	Operational scope
Layer 1	Surveillance data collection and policy constraints initialization	Ethical surveillance policy engine	Input → Urban surveillance streams, ethical policy rules → Output → Contextualized state-action inputs with ethical boundaries
Layer 2	Privacy-constrained policy optimization	Differentially private proximal policy optimization (DP-PPO)	Input → Surveillance actions, state feedback, privacy metrics → Output → Policy decisions regularized for identity protection
Layer 3	Bias mitigation via group fairness learning	Adversarial debiasing constraint	Input → Surveillance action outputs, demographic groups → Output → Penalized policies reducing group disparities
Layer 4	Ethical utility-guided reward calibration	Ethics-aware reward function	Input → Public safety benefit, privacy and bias losses → Output → Aggregated reward balancing utility and harm
Layer 5	Monitoring, audit logging, and ethical oversight	Ethical governance audit module	Input → AI surveillance decisions, compliance rules → Output → Governance reports, fairness logs, and regulatory audit trail

One of the most significant ethical challenges in AI deployment is the opacity of AI systems, often referred to as the “black box problem.” In many cases, the decision-making process within AI algorithms is not easily scrutinized or understood, which raises concerns about their transparency (Sanchez et al. 2024). This lack of transparency becomes especially problematic when AI is used to make real-time decisions about individuals, particularly in public spaces or areas like law enforcement, healthcare, or social services. Without clear insight into how decisions are made, the risk of biased or unfair outcomes increases, disproportionately affecting marginalized communities. For example, these systems might wrongly target or misclassify individuals based on skewed data, which leads to serious ethical dilemmas concerning social justice and equity (Ryan and Gregory 2019).

To address these concerns, it is crucial that AI systems are designed with explainability and accountability at their core. Explainable AI (XAI) techniques, which aim to make AI decision-making more transparent and understandable, are key to ensuring that AI systems can be scrutinized and corrected when necessary. Moreover, ethical frameworks and fairness audits should be integrated into AI development to assess whether these systems align with principles of equity and social justice. Policymakers must ensure that AI systems are transparent and accountable, guaranteeing that their deployment does not perpetuate harm, particularly to vulnerable groups.

To confront the dual challenges of opacity and social bias in AI deployments, we propose a hybrid framework that combines SHapley Additive exPlanations (SHAP) for interpretability with a Group Fairness Disparity Audit Function. This model not only decomposes predictions to show how individual features contribute to decisions but also quantifies whether the outcomes disproportionately affect protected groups. It provides public sector institutions with both local explanation and system-level fairness assessment, ensuring that AI systems used in healthcare, law enforcement, or welfare maintain transparency and uphold social justice standards.

The explainability component is built on SHAP, which decomposes a model prediction $f(x)$ into a sum of feature contributions:

$$f(x) = \phi_0 + \sum_{i=1}^{M} \phi_i$$

where $f(x)$ is the AI model's output, ϕ_0 is the expected model output across the dataset, M is the number of features, and ϕ_i is the Shapley value quantifying the contribution of feature i to the prediction.

To ensure that the predictions are not only explainable but also equitable, we integrate a group disparity audit that quantifies differences in outcome distributions across sensitive subpopulations:

$$\Delta_{fair} = |\mathbb{E}_{x \sim D_A}[f(x)] - \mathbb{E}_{x \sim D_B}[f(x)]|$$

where D_A and D_B represent data distributions from two demographic groups (e.g., majority and minority), $f(x)$ is the AI system's output for input x, and Δ_{fair} measures the magnitude of prediction disparity. A high Δ_{fair} indicates potential algorithmic bias.

When used together, these models provide traceability (via SHAP) and accountability (via Δ_{fair}). They support both local-level explanation (for a single decision) and global fairness audits (across demographic lines), thus operationalizing ethical oversight in complex AI deployments.

This integrated explainability-and-fairness model provides a comprehensive approach to resolving the black box dilemma in AI systems. By leveraging SHAP values, stakeholders can understand exactly which factors influenced each decision. Simultaneously, fairness disparity metrics ensure that AI decisions do not systematically disadvantage vulnerable groups. This framework is particularly critical in public services where opaque algorithms can lead to biased law enforcement profiling, unfair benefit allocation, or unjustified medical prioritization. Embedding this model into the AI lifecycle allows institutions to enforce both interpretability and social accountability advancing a future of transparent, equitable, and responsible AI governance.

Framework: Transparent Accountable Decision Intelligence Framework (TADIF)

Layered AI Architecture for Explainability and Fairness Audits in Public Sector AI

Layer	Main functionality	AI module(s) and components	Operational scope
Layer 1	Sensitive attribute flagging and policy embedding	Fairness policy annotation engine	Input → Dataset metadata and deployment context → Output → Annotated features and fairness policy constraints
Layer 2	Decision explanation via feature attribution	SHAP explainability module	Input → Model predictions, input features → Output → Local decision explanations through feature contributions
Layer 3	Group disparity evaluation for fairness	Group fairness audit metric	Input → Model outputs across demographic groups → Output → Quantified disparity values for protected groups
Layer 4	Audit-based risk reporting and accountability logging	Accountability reporting module	Input → Explanation and fairness results → Output → Interpretability logs and equity audit reports

Furthermore, ethical challenges emerge in the way data is collected, stored, and used. Surveillance technologies often gather vast amounts of data, sometimes without citizens' explicit consent, raising concerns about who controls the data and how it will be used (Ryan and Gregory 2019). The improper use of such data can lead to breaches of personal privacy and can fuel a surveillance state that prioritizes control over individual freedom. In this context, public policies must evolve to ensure that AI-driven systems are held accountable, transparent, and aligned with democratic values.

To formalize this necessity, we introduce a Consent-Aware Data Integrity Function, which ensures that any sample $x \in X$ is ethically valid only if it satisfies jointly the conditions of explicit consent and lawful policy alignment. This constraint becomes a binary-valued gate function that filters all downstream AI processing.

$$\Lambda(x) = \mathbb{I}_{\text{consent}(x)=1} \cdot \mathbb{I}_{\text{policy}(x)=1}$$

where $\Lambda(x)$ is the data legitimacy indicator, $\mathbb{I}$ is the indicator function, $\text{consent}(x)$ returns 1 if explicit consent was given for data sample x, $\text{policy}(x)$ returns 1 if the use of x aligns with data governance regulations. This function filters all data to ensure only ethically admissible samples are used in model inference.

Upon ensuring data compliance, we proceed to the explainability phase, in which we assess the interpretability of a model's prediction using SHAP decomposition. Here, the model output is expressed as a summation over feature-level attributions:

$$f(x) = \varphi_0 + \sum_{i=1}^{M} \varphi_i$$

where $f(x)$ is the AI model's prediction for instance x, φ_0 is the expected model output over the dataset, M is the total number of features, φ_i is the Shapley value of the i-th feature. This function decomposes a decision into its feature contributions, promoting interpretability and accountability.

To ensure these decisions are not only explainable but also socially equitable, we introduce the Group Fairness Disparity Audit, which measures predictive imbalance across sensitive subpopulations. This audit evaluates the absolute difference in mean outcomes between two demographically distinct distributions DADA and DBDB:

$$\Delta_{\text{fair}} = |\mathbb{E}_{x\sim D_A}[f(x)] - \mathbb{E}_{x\sim D_B}[f(x)]|$$

where Δ_{fair} is the fairness disparity metric, $f(x)$ is the model output, $\mathbb{E}_{x\sim D_A}$ is the expected value over group A, $\mathbb{E}_{x\sim D_B}$ is the expected value over group B. This function quantifies group-level inequality in model behavior, detecting demographic bias.

Each of the above formulations is constructed in a way that is mathematically coherent with the prior layer: the ethical legitimacy filter $\Lambda(x)$ constrains what enters the model, the SHAP decomposition $f(x)$ explains the outcome on ethically cleared data, and the disparity audit Δ_{fair} then evaluates systemic fairness across the model's global operation.

This architectural flow integrates consent, transparency, and fairness into a single pipeline ensuring that public sector AI systems align with democratic values and uphold individual dignity at every computational stage.

Framework: Transparent Accountable Decision Intelligence Framework (TADIF)

Layered AI Architecture for Explainability, Fairness, and Consent-Aware Governance in Public Sector AI

Layer	Main functionality	AI module(s) and components	Operational scope
Layer 1	Consent-aware data governance and policy encoding	Consent governance encoding engine	Input → Surveillance-derived personal data → Output → Policy-encoded data with consent metadata and usage constraints
Layer 2	Decision explanation via feature attribution	SHAP explainability module	Input → Model predictions, input features → Output → Local decision explanations through feature contributions
Layer 3	Group disparity evaluation for fairness	Group fairness audit metric	Input → Model outputs across demographic groups → Output → Quantified disparity values for protected groups
Layer 4	Audit-based risk reporting and accountability logging	Ethical accountability audit module	Input → Explanation and fairness results → Output → Interpretability logs and equity audit reports

5.7.1 Public Perceptions and Acceptance of Surveillance

Public perceptions of AI-driven surveillance often revolve around the trade-off between enhanced security and potential privacy violations. Many individuals accept or even welcome increased surveillance if they believe it will contribute to public safety. However, acceptance levels vary widely depending on how the technology is implemented, communicated, and regulated. Public trust is crucial in determining the success of AI surveillance technologies. When surveillance is perceived as intrusive or disproportionately targeting certain groups, it can lead to public backlash and mistrust of governmental institutions (Sanchez et al., 2024). To ensure AI surveillance systems are accepted by the public, it is essential to integrate a mechanism that dynamically learns and adjusts to fluctuating trust levels. We introduce a Bayesian Trust-Adaptive Surveillance Model that allows AI systems to infer public sentiment over time and regulate surveillance behavior accordingly. This model combines probabilistic reasoning with reinforcement learning to encode not only technical performance but also social acceptance metrics into AI decision-making. The model begins by estimating posterior public trust using a Bayesian update mechanism, where prior trust is revised based on observable outcomes:

$$P(\text{Trust}_t|\mathcal{O}_t) = \frac{P(\mathcal{O}_t|\text{Trust}_t) \cdot P(\text{Trust}_{t-1})}{P(\mathcal{O}_t)}$$

where $P(\text{Trust}_t|\mathcal{O}_t)$ is the updated probability of public trust at time t given observations $\mathcal{O}_t$, $P(\mathcal{O}_t|\text{Trust}_t)$ is the likelihood of observing specific societal feedback (e.g., protest signals, sentiment analysis from social media, complaints) given the current trust level, and $P(\text{Trust}_{t-1})$ is the prior belief of trust from the previous cycle.

This probabilistic trust estimation is then embedded into a Trust-Modulated Reward Function within a reinforcement learning framework, enabling the surveillance policy to be adjusted based on trust trends:

$$R(s_t, a_t) = U_{\text{safety}}(a_t) + \theta \cdot T_t - \gamma \cdot C(a_t)$$

where $U_{\text{safety}}(a_t)$ is the public safety utility generated by surveillance action a_t, T_t is the inferred trust level at time t obtained from the Bayesian update, θ is a coefficient translating trust into social capital reinforcement, $C(a_t)$ is the operational or ethical cost associated with the action, and γ balances cost sensitivity. The model thus rewards actions that enhance safety while simultaneously building or maintaining trust. Crucially, when public perception declines e.g., due to disproportionate targeting or lack of transparency the reward diminishes, triggering policy adjustments toward less intrusive or more equitable alternatives.

To ensure long-term adaptability, the model is trained using policy gradient methods with a trust-adjusted learning signal:

$$\nabla_\theta J(\pi_\theta) = \mathbb{E}_t[\nabla_\theta \log \pi_\theta(a_t|s_t) \cdot (R(s_t, a_t) - b_t)]$$

where $\nabla_\theta J(\pi_\theta)$ is the policy gradient, $\log\pi_\theta(a_t|s_t)$ is the log-probability of the current action, and b_t is a baseline function used to reduce variance in learning. This formulation ensures that the system learns to optimize not only for technical performance but also for perceived legitimacy.

This integrated Bayesian-RL model enables AI surveillance systems to monitor and adjust to social feedback in real time. By embedding public trust into both decision-making and training dynamics, it moves AI surveillance from a static enforcement tool to a socially co-regulated governance mechanism. Acceptance is not purely technical; it is political, emotional, and perceptual. Embedding trust as a probabilistic and learnable signal allows cities to preserve security while nurturing democratic legitimacy and civic support in the age of smart surveillance.

Framework: Trust-Adaptive Bayesian Reinforcement Surveillance Framework (TABRSF)

Layered AI Architecture for Public Sentiment-Responsive Surveillance Policy Learning

Layer	Main functionality	AI module(s) and components	Operational scope
Layer 1	Public sentiment collection and policy calibration input	Social feedback ingestion pipeline	Input → Social media data, protests, complaints → Output → Feedback vector for trust estimation and response mapping
Layer 2	Trust estimation via Bayesian inference	Bayesian trust inference module	Input → Prior trust beliefs, social feedback → Output → Posterior probability distribution over public trust levels
Layer 3	Surveillance reward calculation modulated by trust	Trust-adaptive RL reward module	Input → Trust levels, surveillance actions, utility-cost metrics → Output → Adjusted reward values incorporating social perception
Layer 4	Reinforcement learning with trust-sensitive policy optimization	Policy gradient learning module	Input → Trust-modulated rewards and actions → Output → Policy updates optimized for safety and legitimacy
Layer 5	Monitoring public acceptance trends and system retraining	Trust trend monitoring and retraining orchestrator	Input → Long-term feedback and policy performance → Output → Refined trust models and adaptive surveillance strategies

A significant factor influencing public perception is the perceived fairness of AI systems. If AI-driven surveillance technologies are seen as biased or unaccountable, they may erode public confidence in their use. For instance, AI technologies that rely on facial recognition are frequently criticized for racial biases, leading to disproportionate surveillance of ethnic minorities. This perceived inequity can deepen societal divisions and reduce the legitimacy of AI-driven surveillance in the eyes of the public. To address the critical issue of perceived and statistical bias in AI surveillance systems, we propose a Fairness-Constrained Surveillance Risk Model grounded in Statistical Parity Difference (SPD). This model is designed to ensure that AI-powered systems, particularly those relying on facial recognition or behavioral profiling, do not disproportionately affect certain demographic groups. It quantifies fairness disparities in detection or flagging rates across protected attributes (e.g., race, gender), offering both a mathematical assessment and a constraint mechanism for mitigation.

The fairness disparity is formally measured by the Statistical Parity Difference:

$$\Delta_{SPD} = |Pr(\hat{Y} = 1|Z = 0) - Pr(\hat{Y} = 1|Z = 1)|$$

where $\hat{Y} = 1$ denotes a positive surveillance action (e.g., flagged for further investigation), Z is the binary protected attribute (e.g., 0 = majority group, 1 = minority group), and Δ_{SPD} measures the absolute difference in positive decision rates across demographic groups.

To minimize the risk of perceived unfairness, a constrained surveillance loss function can be introduced:

$$\underset{\theta}{min}\mathbb{E}_{(x,y)}[L(f_\theta(x), y)]\text{subject to } \Delta_{SPD} \leq \delta$$

where$f_\theta(x)$ is the prediction of the AI surveillance model, $\mathcal{L}(\cdot)$ is the classification loss (e.g., binary cross-entropy), and δ is a maximum allowable disparity threshold established by policy. This constraint ensures that the trained model minimizes predictive loss while maintaining acceptable parity in treatment across groups.

To monitor real-time compliance, the disparity measure can be embedded into a public transparency dashboard, reinforcing institutional accountability and enhancing public trust.

This fairness-constrained risk model quantifies and limits demographic disparities in AI surveillance outputs. By operationalizing fairness through Statistical Parity Difference and embedding it as a constraint in model optimization, public institutions can prevent algorithmic surveillance systems from disproportionately targeting minority populations. The model strengthens the perceived legitimacy and ethical standing of AI by aligning technical performance with societal values of equity and nondiscrimination. When deployed in real-time monitoring systems such as facial recognition for public safety or smart city surveillance, this framework ensures that fairness is not an afterthought, but a rigorously enforced design principle.

Framework: Fairness-Constrained Surveillance Risk Framework (FCSRF)

Layered AI Architecture for Bias-Reduced and Legitimacy-Promoting AI Surveillance

Layer	Main functionality	AI module(s) and components	Operational scope
Layer 1	Preprocessing of surveillance data with bias filtering	Bias-aware data preprocessing engine	Input → Raw surveillance data → Output → Filtered data for fairness-compliant processing

(continued)

(continued)

Layer	Main functionality	AI module(s) and components	Operational scope
Layer 2	Fairness evaluation via statistical parity difference	Statistical parity difference metric module	Input → Model outputs, demographic data → Output → Quantified fairness disparity metric across groups
Layer 3	Fairness-constrained policy optimization	Fairness-constrained learning module	Input → SPD measure, model predictions → Output → Optimized model trained under fairness constraints
Layer 4	Public transparency and ethical oversight	Fairness transparency audit module	Input → Model performance logs, SPD metrics → Output → Transparency dashboard and fairness audit reports

The public's acceptance of surveillance also depends on the transparency and accountability of the systems in place. When citizens are adequately informed about how their data is being used and when there are robust measures to prevent misuse, there tends to be greater acceptance. Engaging communities in discussions about surveillance technologies, ensuring that data collection aligns with privacy laws, and implementing clear oversight mechanisms are critical to gaining public trust (Sanchez et al. 2024). To address this challenge, we incorporate data compliance and ethical governance mechanisms at the very beginning of the surveillance system pipeline. This is achieved through a Compliance Filter Module, which ensures that only legally compliant data is fed into the system for further processing. The filter can be formally defined using the following equation, ensuring that only ethically sound and legally obtained data enters the surveillance system.

$$\Lambda_{\text{privacy}}(x) = \mathbb{I}_{\text{compliant}(x)=1} \cdot \mathbb{I}_{\text{consent}(x)=1}$$

where $\Lambda_{\text{privacy}}(x)$ is the privacy filter for data x, $\mathbb{I}$ is the indicator function, compliant(x) returns 1 if data sample x complies with privacy laws (e.g., GDPR), consent(x) returns 1 if the individual provided explicit consent. This function ensures that data entering the system is compliant with privacy laws and consent protocols before it is processed.

Following data compliance, we focus on ensuring system transparency. To achieve this, an Engagement and Oversight Module is incorporated, which captures real-time public feedback regarding surveillance operations. This feedback allows the AI system to adjust its decisions in real time, ensuring that the system remains transparent and responsive to public sentiment. This dynamic adaptation of trust is modeled using a sentiment feedback mechanism, where public sentiment S_t evolves based on feedback F_t:

$$S_t = \alpha \cdot S_{t-1} + (1 - \alpha) \cdot F_t$$

where S_t is the sentiment score at time t, S_{t-1} is the previous sentiment score, F_t is the real-time feedback from the public, and α is the smoothing factor that controls how quickly the system adapts to new feedback. This function ensures that the system's behavior is continuously aligned with public sentiment and that the surveillance system remains accountable and transparent.

After incorporating public feedback mechanisms, the next key consideration is to ensure the fairness of the surveillance system. This is achieved through the integration of a fairness-constrained optimization approach. The fairness disparity is formally measured by the Statistical Parity Difference (SPD), which quantifies the difference in surveillance treatment across protected demographic groups:

$$\Delta_{\text{SPD}} = |Pr(\hat{Y} = 1 \mid Z = 0) - Pr(\hat{Y} = 1 \mid Z = 1)|$$

where Δ_{SPD} is the Statistical Parity Difference, which quantifies the absolute difference in decision outcomes between two groups, $\Pr(\hat{Y} = 1 \mid Z = 0)$ is the probability that the model gives a positive decision (e.g., "flagged for further investigation") given that the individual belongs to the majority group $Z = 0$ (e.g., non-minority group), $\Pr(\hat{Y} = 1 \mid Z = 1)$ is the probability that the model gives a positive decision given that the individual belongs to the minority group $Z = 1$ (e.g., minority group). The formula measures the absolute difference in these probabilities, i.e., how much more or less likely the model is to flag someone from one group compared to another.

To minimize the fairness risk, a constrained optimization approach is applied in which the model's objective function is penalized when disparities exceed a given threshold. This is formulated as:

$$\underset{\theta}{min}\mathbb{E}_{(x,y)}\left[L(f_\theta(x), y)\right]\text{s.t. } \Delta_{\text{SPD}} \leq \delta$$

where $f_\theta(x)$ is the prediction of the surveillance model, $L(\cdot)$ is the classification loss function, and δ is a maximum allowable disparity threshold established by policy. This constraint ensures that the trained model minimizes predictive loss while maintaining fairness across groups.

This framework ensures that surveillance systems are transparent, accountable, and fair, while respecting privacy laws and maintaining public trust. By embedding these fairness and transparency modules directly into the system's learning and operational processes, it is possible to create an AI-driven surveillance infrastructure that upholds ethical standards and democratic values in real time.

Framework: Fairness-Constrained Surveillance Risk Framework (FCSRF)

Layered AI Architecture for Privacy-Compliant, Transparent, and Accountable Surveillance Systems

Layer	Main functionality	AI module(s) and components	Operational scope
Layer 1	Data compliance and consent verification	Consent compliance validation engine	Input → Raw surveillance data → Output → Privacy-compliant data with consent metadata and legal checks

(continued)

(continued)

Layer	Main functionality	AI module(s) and components	Operational scope
Layer 2	Real-time public sentiment feedback integration	Sentiment feedback module	Input → Public feedback, sentiment data → Output → Dynamic adjustment of surveillance behavior based on public trust
Layer 3	Fairness evaluation via statistical parity difference	Statistical parity difference metric module	Input → Surveillance outputs, demographic data → Output → Quantified fairness disparity across groups
Layer 4	Fairness-Constrained Optimization of Surveillance Actions	Fairness-constrained learning module	Input → SPD metric, model predictions → Output → Optimized surveillance policy under fairness constraints
Layer 5	Real-time transparency monitoring and ethical oversight	Ethical transparency governance module	Input → Surveillance logs, fairness metrics → Output → Transparency dashboard with fairness audit reports

Figure 5.2 explores the ethical and social dimensions of AI-driven surveillance through three interconnected areas linked to the central node, "AI-Driven Surveillance": ethical concerns, data collection and usage, and public perceptions and acceptance. Ethical concerns include privacy risks from constant monitoring, such as facial recognition and IoT data collection, "black box" opacity and lack of transparency in AI decisions, and potential bias that may disproportionately affect marginalized communities. Data collection and usage focus on concerns over data collection without consent, risks of data misuse, impacts on individual freedom, and the need for accountable, transparent, and democratic-aligned policies. Public perceptions and acceptance address the trade-off between security benefits and privacy violations, the influence of perceived fairness and bias on public trust, and the importance of transparency, accountability, and community engagement in building trust. Overall, the figure presents AI-driven surveillance as a domain shaped by privacy, consent, transparency, fairness, democratic accountability, and public acceptance.

AI-driven surveillance in urban areas presents significant ethical challenges, particularly regarding privacy, equity, and transparency. As cities become more reliant on AI to manage public spaces, it is essential to address these ethical concerns through public engagement, transparent policies, and robust oversight mechanisms. Public perceptions of AI surveillance are influenced by trust in the system's fairness and the perceived balance between security and personal privacy.

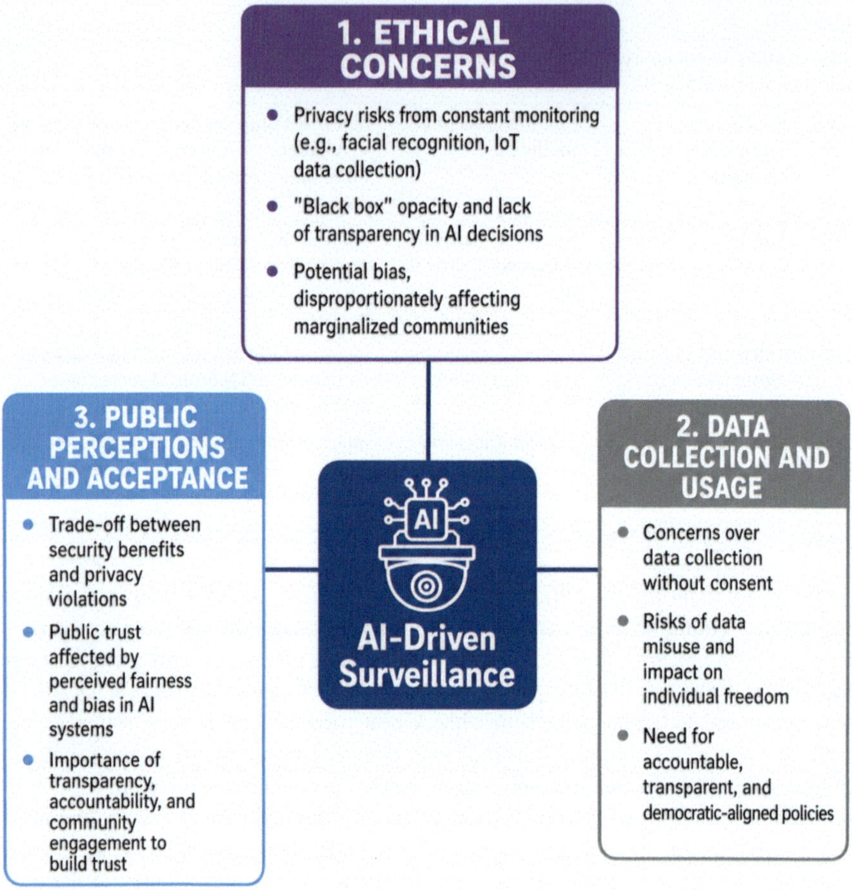

Fig. 5.2 Ethical and social dimensions of AI-driven surveillance

5.8 Conclusion

This chapter has demonstrated that securing AI-driven public services is not simply a technical exercise but a civic necessity. It has shown that without end-to-end protections, every layer of a digital government pipeline becomes a point of vulnerability whether through unencrypted data flows, poorly controlled access, adversarial manipulation of models, or the silent leakage of private information. Each of these weaknesses threatens not only operational continuity but also the trust that underpins democratic legitimacy.

The chapter advances the argument that resilience must be engineered into the architecture itself. Encryption, privacy-preserving computation, and consent-aware data governance provide a baseline of confidentiality. Continuous auditing, adversarial training, and robustness evaluation transform fragile models into systems capable of withstanding targeted attacks and unexpected shocks. Differential privacy and anonymization ensure that analytics can scale without endangering individuals, while decision-theoretic response strategies formalize how institutions contain harm when incidents occur. Together, these elements shift the paradigm from reactive patchwork defenses to proactive, standards-ready security ecosystems.

Most importantly, the chapter insists that ethics and accountability must remain inseparable from technical security. By embedding transparency, fairness, and citizen oversight into the very same frameworks that enforce privacy and robustness, AI in public services can be made not only efficient but also equitable and trustworthy. The overarching lesson is clear: a secure AI infrastructure is the foundation upon which digital governance, citizen confidence, and societal resilience depend.

References

Ryan, M., & Gregory, A. (2019). Ethics of using smart city AI and big data: The case of four large European cities. *Urban Science*. https://doi.org/10.29297/orbit.v2i2.110

Sanchez, T. W., Brenman, M., & Ye, X. (2024). The ethical concerns of artificial intelligence in urban planning. *Journal of Urban Planning and Development*. https://doi.org/10.1080/01944363.2024.2355305

Sangwan, R.S., Badr, Y., & Srinivasan, S.M. (2023). Cybersecurity for AI Systems: A Survey. *Journal of Cybersecurity and Privacy*. https://doi.org/10.3390/jcp3020010

Villegas-Ch, W., & García-Ortiz, J. (2023). Toward a Comprehensive Framework for Ensuring Security and Privacy in Artificial Intelligence. *Electronics*. https://doi.org/10.3390/electronics12183786

Chapter 6
Artificial Intelligence and Social Inclusion: Designing Equitable Smart Infrastructures

This chapter addresses a central promise and peril of the intelligent city: whether AI amplifies opportunity or calcifies exclusion depends on how we design for inclusion from the start. It develops a rigorous, deployment-ready architecture that detects who is being left behind, reallocates resources in real time to close gaps, and encodes equity as a constraint rather than a hope. The narrative begins with forward-looking risk sensing: sequence models estimate the probability of digital exclusion across groups and places, turning signals like connectivity, device access, literacy, disability, age, income, and geography into actionable forecasts. On top of prediction, a learning policy reallocates interventions: public Wi-Fi hubs, subsidized devices, training programs in real time, optimizing a social-inclusion reward that balances risk reduction, accessibility gains, and budget.

Beyond speed and scale, the chapter formalizes fairness. An inclusion-aware utility with minimum-service thresholds guarantees that no demographic or region falls below a baseline of access, while a geospatial equity audit surfaces rural and peri-urban blind spots that often escape aggregate statistics. Human–AI interaction is treated as a first-class equity problem: a multi-objective interface optimization minimizes cognitive load, closes performance gaps across user groups, and adapts language, modality, and layout to diverse abilities. Affordability moves from rhetoric to design via subsidy optimization and public-hub siting, integrated directly into the global objective so cost constraints never masquerade as inevitabilities. Finally, democratic preference aggregation transforms participation into computation, converting citizen feedback into weights that steer policies without sacrificing representational balance.

This chapter offers a blueprint for precision inclusion at population scale predictive where it must be, normative where it should be, and practical where it counts so that AI-enabled infrastructures expand capability, dignity, and access for all, not just the already connected.

D. Sargiotis, *Empathy in Algorithms*, https://doi.org/10.1007/978-3-032-32128-2_6

6.1 Ensuring AI and Smart Infrastructure Benefits Are Accessible to All

In the context of smart cities and AI-driven infrastructures, barriers to access for marginalized communities can be multifaceted. Digital divides represent a significant hurdle for many groups. As explored by Colding et al. (2024), first and second-level digital divides concerning access to digital tools and the ability to use them effectively limit the ability of marginalized groups to engage with and benefit from smart city infrastructures. For instance, individuals with lower socio-economic status may lack access to high-speed internet or smart devices, placing them at a disadvantage in a digitally-driven society. Furthermore, these individuals might also face difficulties in using the technology due to insufficient digital literacy skills (Colding et al. 2024). To ensure that AI and smart infrastructure systems promote true social inclusion, we introduce a two-tiered AI-driven framework: the first component uses a Long Short-Term Memory (LSTM) model to forecast the probability of exclusion for various social groups over time; the second component applies Reinforcement Learning (RL) to dynamically reallocate smart infrastructure resources in response to these forecasts. This dual architecture enables both early detection of digital inequity and real-time corrective interventions. The Exclusion Risk Prediction Function using LSTM is defined as:

$$h_t = f_{\text{LSTM}}(x_t, h_{t-1}) \text{ and } \hat{y}_t = \sigma(Wh_t + b)$$

where x_t is the input vector at time t representing features such as internet access, device ownership, digital literacy level, and income bracket; h_t is the hidden state capturing temporal exclusion dynamics; $f_{\text{LSTM}}(\cdot)$ is the LSTM transition function; W and b are the weight matrix and bias; and $\hat{y}_t$ is the predicted probability of digital exclusion at time t. The activation function σ ensures predictions are normalized between 0 and 1. Once exclusion risk is estimated for each demographic cluster, the system invokes a policy $\pi(a_t|s_t)$ to deploy digital inclusion resources e.g., community Wi-Fi hubs, digital literacy campaigns, or subsidized devices. This policy is learned via reinforcement learning, where decisions are guided by a Social Inclusion Reward Function:

$$R(s_t, a_t) = -\alpha \cdot \hat{y}_t + \beta \cdot A(a_t) - \gamma \cdot C(a_t)$$

where $\hat{y}_t$ is the predicted exclusion risk (penalized), $A(a_t)$ is the accessibility gain resulting from action a_t (e.g., increase in broadband users or literacy rates), $C(a_t)$ is the resource cost of implementing a_t, and α, β, γ are tuning parameters controlling sensitivity to risk, impact, and cost.

Finally, the policy is optimized using a standard RL objective:

$$\pi^* = \arg\max_{\pi} \mathbb{E}_t[R(s_t, a_t)]$$

This function ensures that the learned policy maximizes long-term accessibility while minimizing exclusion risk and respecting infrastructure budgets.

This dual-model system addresses both predictive and prescriptive challenges of social inclusion. By forecasting digital exclusion using LSTM, the framework provides a forward-looking signal that guides smart infrastructure deployment. The integration of reinforcement learning ensures that interventions are both responsive and cost-efficient. Together, these components enable governments and urban planners to evolve their systems from passive equity compliance to active digital empowerment where smart infrastructure is not just present, but meaningfully accessible to all.

Framework: AI-Driven Digital Inclusion and Resource Allocation Framework (AIDIRAF)
Layered AI Architecture for Predictive Exclusion Risk and Resource Reallocation in Smart Cities

Layer	Main functionality	AI module(s) and components	Operational scope
Layer 1	Preprocessing data for inclusion risk prediction	Inclusive data preprocessing engine	Input → Raw demographic and socio-economic data → Output → Fair data for further processing
Layer 2	Exclusion risk forecasting via LSTM	Long short-term memory (LSTM) network for time-series prediction	Input → Demographic data, digital access indicators → Output → Predicted exclusion risk over time
Layer 3	Social inclusion resource allocation	Custom reward function for social inclusion in reinforcement learning	Input → Exclusion risk predictions, community feedback → Output → Actionable resource deployment strategies
Layer 4	Reinforcement learning for policy optimization	Policy optimization in reinforcement learning	Input → Social inclusion rewards, action outcomes → Output → Optimized policy for dynamic resource reallocation
Layer 5	Monitoring, public feedback, and adaptation	Adaptive inclusion feedback module	Input → Real-time system performance data, community feedback → Output → Adaptive policies for continued digital inclusion

Barriers also include social exclusion where certain populations, such as elderly people or people with disabilities, are left out from actively participating in smart infrastructures because the systems are not designed with their specific needs in mind (Malek et al. 2021). This exclusion happens when technology is designed with the assumption of a "tech-savvy" user base, inadvertently excluding those who may not have the necessary digital skills or access to technological resources. Additionally, geographical disparities play a role. In rural or underdeveloped areas, infrastructure may not be robust enough to support AI-based smart systems (Malek et al. 2021).

To address the structural exclusion of marginalized groups in AI-powered smart infrastructures, we propose an Inclusion-Aware Utility Optimization Model constrained by accessibility equity conditions. This model quantitatively evaluates AI system performance across different user groups and geographic areas, ensuring that smart systems are designed and deployed to serve not only tech-literate populations in urban centers but also digitally underserved communities, elderly users, and persons with disabilities.

At the core of this framework is a demographically weighted utility function:

$$U_{total} = \sum_{g=1}^{G} w_g \cdot \mathbb{E}_{x \sim D_g}[U_g(f(x))]$$

where G is the number of demographic or geographic groups (e.g., urban users, rural residents, elderly, disabled), w_g is the societal weight or policy priority for group g, $U_g(f(x))$ is the group-specific utility function measuring benefit from system output.

To enforce equity, the model introduces a minimum accessibility constraint:

$$\mathbb{E}_{x \sim D_g}[U_g(f(x))] \geq \eta, \forall g \in G$$

where η is a predefined minimum utility threshold that must be satisfied for all population segments. This constraint ensures that no group falls below a basic level of service or inclusion.

To identify underserved areas, a geospatial equity audit function can be defined as:

$$\Delta_{geo} = \max_{r_1, r_2 \in R} |\mathbb{E}_{x \sim D_{r_1}}[U(f(x))] - \mathbb{E}_{x \sim D_{r_2}}[U(f(x))]|$$

where R is the set of all regions (e.g., rural, suburban, urban), and Δ_{geo} measures the worst-case spatial disparity in utility. Minimizing Δ_{geo} ensures geographic fairness in access and experience.

This inclusion-aware optimization model empowers policymakers and system designers to embed fairness, accessibility, and geographic equity directly into AI-driven smart infrastructure. It formalizes the commitment to serve diverse and vulnerable populations, ensuring that elderly users, people with disabilities, and rural residents are not inadvertently excluded due to system assumptions or digital infrastructure gaps. By mathematically encoding inclusion thresholds and utility disparities, this framework ensures that smart city systems evolve in a socially equitable, universally accessible, and ethically aligned manner.

Framework: Inclusion-Aware Utility Optimization Framework (IAUOF)
Layered AI Architecture for Digital Inclusion and Geographic Equity in Smart Cities

Layer	Main functionality	AI module(s) and components	Operational scope
Layer 1	Data preprocessing and demographic grouping	Demographic data normalization engine	Input → Raw socio-economic and demographic data → Output → Fair data for further processing
Layer 2	Utility calculation for each group using weighted aggregation	Demographic-aware utility aggregation	Input → Demographic data, utility functions → Output → Aggregated utility across groups
Layer 3	Equity enforcement through minimum utility threshold	Equity-constrained optimization	Input → Utility values, predefined thresholds → Output → Optimized system design with equity constraints
Layer 4	Geospatial equity analysis and resource distribution	Spatial fairness assessment	Input → Geospatial data, utility values → Output → Resource allocation ensuring spatial fairness
Layer 5	Monitoring and real-time feedback for continuous adaptation	Adaptive equity feedback module	Input → Real-time system performance, community feedback → Output → Adaptive smart city policies

6.2 Strategies for Inclusivity in the Design of AI Systems

To ensure inclusivity, it is crucial to design AI systems with universal accessibility in mind, as emphasized by Malek et al. (2021). AI-driven infrastructure should be adaptable to the varying needs of different groups, particularly those from marginalized communities. For example, integrating multi-language support and easy-to-use interfaces can make AI systems more accessible to non-native speakers and individuals with lower digital literacy levels (Malek et al. 2021). To systematically support inclusivity at the design stage, we propose an Adaptive Interface Optimization Model that integrates multi-objective machine learning with constraints derived from universal design principles. This model enables AI systems to learn interface adaptations that maximize accessibility across diverse user profiles, including non-native speakers, elderly users, individuals with low digital literacy, and people with cognitive or motor impairments. The core of the model is a multi-objective optimization function:

$$\min_{\theta}[\mathcal{L}_{\text{task}}(\theta) + \lambda_1 \cdot \mathcal{L}_{\text{usability}}(\theta) + \lambda_2 \cdot \mathcal{L}_{\text{accessibility}}(\theta)]$$

where θ represents the AI system parameters (e.g., language model weights, UI configuration), $\mathcal{L}_{\text{task}}(\theta)$ is the primary task performance loss (e.g., accuracy in response generation), $\mathcal{L}_{\text{usability}}(\theta)$ quantifies cognitive load and interface complexity across user groups, $\mathcal{L}_{\text{accessibility}}(\theta)$ measures inclusion deficits (e.g., language mismatch, visual overload), and λ_1, λ_2 are weighting factors that balance efficiency with inclusivity.

To model usability and accessibility in diverse populations, we define $\mathcal{L}_{\text{usability}}(\theta)$ as:

$$\mathcal{L}_{\text{usability}}(\theta) = \sum_{u \in U} w_u \cdot \text{CognitiveLoad}(u, \theta)$$

where U is the set of user subgroups, w_u is a demographic weight based on user prevalence or priority (e.g., elderly, low-literacy users), and $\text{CognitiveLoad}(u, \theta)$ is the system-estimated effort for user group uu to complete a given task using interface configuration θ.

Simultaneously, the Accessibility Loss is defined using a fairness-oriented adversarial term:

$$\mathcal{L}_{\text{accessibility}}(\theta) = \underset{d \in D}{max} |\mu^{(d)}(\theta) - \overline{\mu}(\theta)|$$

where D is the set of demographic groups, $\mu^{(d)}(\theta)$ is the average performance or satisfaction score for group d, and $\overline{\mu}(\theta)$ is the global average. This term penalizes interface configurations that disproportionately disadvantage certain user groups, reinforcing the principle of fairness in interaction.

These functions are optimized jointly during system training or interface tuning, using stochastic gradient descent with demographic feedback loops.

This AI-driven design optimization model empowers developers to embed inclusion constraints directly into the system architecture rather than retrofitting accessibility later. By quantifying and minimizing interface complexity, cognitive strain, and demographic performance gaps, the system actively learns how to adapt itself to the real-world needs of all users. As emphasized by Malek et al. (2021), true accessibility is not a feature layer—it is a foundational design logic. This framework ensures that smart infrastructures evolve not only in sophistication but in fairness, usability, and human dignity.

Framework: Adaptive Interface Optimization Framework (AIOF)
Layered AI Architecture for Usability, Accessibility, and Equity in Smart Infrastructure

Layer	Main functionality	AI module(s) and components	Operational scope
Layer 1	Preprocessing data for accessibility and usability	Accessibility data preprocessing engine	Input → Raw demographic and usage data → Output → Fair and usable data for further processing
Layer 2	Utility calculation for each group using weighted aggregation	Adaptive interface optimization	Input → Demographic data, system outputs → Output → Group-specific utility function

(continued)

(continued)

Layer	Main functionality	AI module(s) and components	Operational scope
Layer 3	Equity enforcement through minimum accessibility threshold	Equity-constrained optimization	Input → Utility values, predefined thresholds → Output → System design optimized under fairness constraints
Layer 4	Geospatial equity analysis and interface adaptation	Spatial fairness assessment	Input → Geospatial data, utility values → Output → Fair resource allocation ensuring geographic inclusion
Layer 5	Monitoring and real-time feedback for continuous adaptation	Adaptive usability feedback module	Input → Real-time user feedback, system performance data → Output → Adaptive policies for equitable system design

Another strategy includes the implementation of affordability measures for accessing AI services and smart infrastructure. Governments and public institutions should ensure that necessary digital infrastructure is made available at lower costs, particularly in underprivileged areas. By creating public digital hubs with free or subsidized internet access and AI tools, marginalized communities can better engage with smart city technologies (Colding et al. 2024).

To integrate affordability into the adaptive AI framework previously discussed, we introduce a Subsidy-Based Resource Allocation Model, which optimizes the distribution of subsidized services (e.g., internet access and AI tools) based on regional needs. The cost for providing these services is calculated using the following equation:

$$C_{\text{subsidy}} = \sum_{r \in R} [\text{Subsidy Rate}(r) \cdot \text{Demand}(r)]$$

where C_{subsidy} is the total cost for providing subsidized services across all regions, Subsidy Rate(r) represents the percentage of the subsidy provided to residents in region r, for instance, 50% off the service cost, Demand(r) quantifies the demand for AI tools and internet services in region r, which is driven by population density and socio-economic factors. This function computes the total cost of providing the necessary subsidies based on the demand in each area, ensuring equitable access for underserved communities.

Next, we introduce an optimization function for determining the optimal locations of public digital hubs that provide subsidized services. The goal is to maximize the accessibility of these hubs in areas where demand for digital resources is highest. This is formulated as:

$$H_{\text{opt}} = \arg \underset{h \in H}{max} [\text{Access Need}(h) \cdot \text{Community Demand}(h)]$$

Here: H_{opt} represents the optimal locations for public digital hubs, Access Need(h) quantifies how much infrastructure is needed in location h, such as rural or underserved areas, Community Demand(h) measures the level of demand for AI tools and internet access within the community in location hh. This function helps identify and allocate digital hubs strategically, focusing on areas with the highest need for affordable digital access.

To ensure that these affordability measures align with the overall smart city infrastructure design, we embed the cost-accessibility considerations into the previous framework's utility maximization function. This integration helps ensure that cost-effectiveness is balanced with the need for universal accessibility. The extended multi-objective optimization now includes the subsidy and access costs:

$$\min_{\theta}[L_{\text{task}}(\theta) + \lambda_1 \cdot L_{\text{usability}}(\theta) + \lambda_2 \cdot L_{\text{accessibility}}(\theta) + \lambda_3 \cdot C_{\text{subsidy}}]$$

where C_{subsidy} is the total cost of subsidies computed from the previous equation, λ_3 is the weight factor that balances the cost-effectiveness of subsidies with usability and accessibility in the optimization process. This updated function ensures that the system not only adapts to user needs but also integrates cost-effective and equitable strategies for providing AI-powered infrastructure.

By incorporating these affordability measures into the framework, we ensure that AI systems are not only accessible but also affordable, making sure that marginalized communities are included in the digital transformation of smart cities. This approach aligns technical performance with social inclusion and ensures that universal design principles are embedded throughout the system's lifecycle.

Framework: Adaptive Interface Optimization Framework with Affordability Measures (AIOFAM)

Layered AI Architecture for Usability, Accessibility, Affordability, and Equity in Smart Infrastructure

Layer	Main functionality	AI module(s) and components	Operational scope
Layer 1	Preprocessing data for affordability and accessibility	Affordability data preprocessing engine	Input → Raw socio-economic, digital resource, and demographic data → Output → Fair and usable data for further processing
Layer 2	Cost-effective resource allocation and subsidy calculation	Subsidy-based resource allocation model	Input → Demographic data, system outputs → Output → Cost computation for subsidies across regions
Layer 3	Optimal location identification for public digital Hubs	Public digital hubs optimization	Input → Geospatial data, community demand → Output → Identified locations for digital hubs based on needs
Layer 4	Equity and accessibility enforcement in system design	Adaptive interface optimization with cost-accessibility integration	Input → Subsidy cost, interface configuration, usability/ accessibility metrics → Output → Optimized system design that balances cost and accessibility

(continued)

(continued)

Layer	Main functionality	AI module(s) and components	Operational scope
Layer 5	Monitoring and real-time feedback for continued adaptation	Equity adaptation feedback module	Input → Real-time system feedback, user experience data → Output → Continuous adaptation and policy adjustments for equity and accessibility

Citizen engagement is another critical component. Inclusivity efforts must include soliciting input from marginalized communities to understand their unique needs, ensuring that these voices are heard in the design and development phases of AI systems. Colding et al. (2024) emphasize that the participation of citizens in decision-making processes can help reduce disparities in access and usage of smart technologies. To operationalize the inclusion of marginalized communities in AI system design, we propose a Democratic Preference Aggregation Framework that embeds stakeholder feedback directly into the optimization of AI objectives. This model ensures that diverse citizen inputs are not only collected but mathematically encoded into the utility function of the AI system, allowing underrepresented voices to guide both policy-level and system-level decisions. It formalizes participatory AI design by translating qualitative input into quantifiable preference weights that influence system training, model selection, or resource allocation in smart infrastructures.

At the core of this framework is the weighted participatory utility objective:

$$U_{collective} = \sum_{g=1}^{G} \pi_g \cdot \mathbb{E}_{x \sim D_g}[U_g(f(x))]$$

where G is the number of community stakeholder groups (e.g., low-income residents, elderly, disabled, urban/rural populations), π_g is the participatory weight assigned to group g based on citizen engagement inputs, D_g is the distribution of inputs from group g, and $U_g(f(x))$ is the perceived utility of system output $f(x)$ for that group.

The participatory weights πgπg are derived from a normalized feedback aggregation function based on community consultation data:

$$\pi_g = \frac{\phi_g}{\sum_{k=1}^{G} \phi_k}$$

where ϕ_g is the aggregated importance or priority level for group g, as determined through participatory workshops, surveys, or digital deliberation platforms. This ensures that the collective utility function aligns with the actual preferences and concerns expressed by citizens during engagement processes.

To maintain representational fairness, a participation balance constraint can be enforced:

$$|\pi_g - \frac{n_g}{N}| \leq \delta, \forall g$$

where n_g is the population count of group gg, $N = \sum_g n_g$ is the total population, and δ is a fairness tolerance threshold. This constraint ensures that participation weights do not diverge unreasonably from demographic proportions unless justified by engagement results.

This citizen engagement framework ensures that public AI systems are co-designed with the communities they intend to serve. By embedding stakeholder preferences directly into the utility optimization process, it transforms participation from symbolic consultation into computational influence. This model supports transparent civic collaboration, reduces exclusionary design risks, and ensures that marginalized voices actively shape smart technology outcomes. In line with Colding et al. (2024), it institutionalizes inclusive decision-making in AI infrastructure planning advancing not only democratic governance but algorithmic equity.

Framework: Democratic Preference Aggregation Framework (DPAF)
Layered AI Architecture for Inclusive Citizen Engagement and Fair AI Systems

Layer	Main functionality	AI module(s) and components	Operational scope
Layer 1	Data collection and preprocessing for participatory inputs	Participatory data preprocessing engine	Input → Raw citizen input data (surveys, workshops) → Output → Processed data for further analysis
Layer 2	Utility calculation and weighted aggregation	Weighted participatory utility aggregation	Input → Demographic and utility data, participatory weights → Output → Collective utility function based on citizen inputs
Layer 3	Normalization of feedback aggregation and utility function alignment	Normalized feedback aggregation function	Input → Aggregated importance scores from citizen inputs → Output → Normalized weights for each stakeholder group
Layer 4	Fairness enforcement through participation balance constraint	Participation balance constraint	Input → Population data, participation weights → Output → Constraints ensuring equitable participation across groups
Layer 5	Policy optimization and real-time adaptation	Inclusive policy adaptation module	Input → Real-time feedback, utility values → Output → Optimized policies ensuring inclusive decision-making

Lastly, continuous education and training programs are essential to bridge the digital literacy gap. Programs tailored to enhance the skills of underrepresented groups, such as older adults or low-income individuals, can significantly improve their ability to navigate and utilize AI-driven infrastructures (Malek et al. 2021).

Figure 6.1 illustrates key aspects of promoting social inclusion within AI and smart infrastructure systems, divided into barriers to access and strategies for inclusivity. Barriers include digital divides, where marginalized groups may lack access to digital tools or possess limited digital literacy; social exclusion, which affects groups

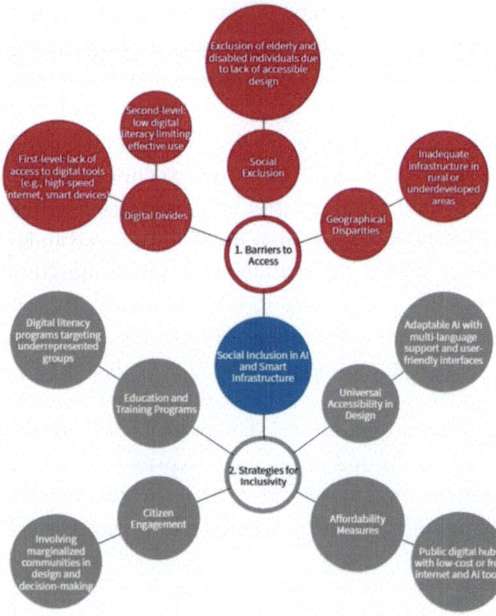

Fig. 6.1 Social inclusion in AI and smart infrastructure

like the elderly and disabled due to inaccessible design; and geographical disparities, where rural or underdeveloped areas may lack robust infrastructure. Strategies for inclusivity involve designing universally accessible AI with multi-language support, implementing affordability measures such as public digital hubs, fostering citizen engagement by involving marginalized communities in decision-making, and providing education and training programs to improve digital literacy among underrepresented groups. Each area is color-coded to distinguish between the barriers and strategies, providing a comprehensive overview of how to ensure AI and smart infrastructure are accessible to all.

To achieve true inclusivity in AI and smart infrastructure systems, it is essential to address both technological and social barriers. By focusing on affordability, accessibility, education, and inclusive decision-making processes, policymakers and technologists can ensure that the benefits of AI and smart infrastructure are accessible to all members of society, especially marginalized groups.

6.3 Conclusion

This chapter has demonstrated that social inclusion in AI-driven smart infrastructures cannot be left as an aspiration; it must be engineered as a measurable and enforceable outcome. It has shown that digital divides, social exclusion, and geographic disparities are not peripheral issues but structural risks that, if unaddressed, can turn intelligent cities into engines of inequality. By contrast, when inclusion is treated as a design principle, AI can actively identify exclusion risks, reallocate resources dynamically, and embed fairness into the very logic of governance.

The frameworks developed in this chapter translate principles into practice. Forecasting models anticipate who is at risk of being left behind, while reinforcement learning dynamically allocates interventions to minimize inequity. Utility optimization functions ensure that every group—urban or rural, elderly or digitally skilled—meets minimum thresholds of access and dignity. Adaptive interfaces and affordability models prove that inclusivity is not an afterthought but a quantifiable dimension of system performance. Participatory frameworks elevate citizen voices from symbolic consultation to computational influence, ensuring that policies evolve in line with diverse social priorities.

The overarching lesson is clear: intelligent infrastructures that ignore equity will magnify exclusion, but those that embed inclusion at every layer can transform digital transformation into social empowerment. The chapter offers a pragmatic blueprint for governments and city leaders worldwide predictive in foresight, normative in principle, and practical in application to ensure that the benefits of AI extend universally, expanding capability, dignity, and trust across all communities.

References

Colding, J., Nilsson, C., & Sjöberg, S. (2024). Smart Cities for All? Bridging Digital Divides for Socially Sustainable and Inclusive Cities. *Smart Cities*, *7*(3), 1044–1059. https://doi.org/10.3390/smartcities7030044

Malek, J. A., Lim, S. B., & Yigitcanlar, T. (2021). Social Inclusion Indicators for Building Citizen-Centric Smart Cities: A Systematic Literature Review. *Sustainability*, *13*(1), 376. https://doi.org/10.3390/su13010376

Chapter 7
AI and the Digital Divide: Building Equitable Smart Cities

This chapter confronts the urban digital divide with a single proposition: inclusion must be measurable, optimizable, and governed not merely proclaimed. It assembles a deployment-ready architecture that cities can use to diagnose inequities precisely, target interventions dynamically, and guarantee minimum levels of digital capability for every community. The narrative begins with measurement: a Digital Equity Index fuses access to devices, broadband quality, and skills into a single actionable metric, then enforces equity floors so no demographic or neighborhood falls below a basic standard of service.

Diagnosis turns spatial. Graph Neural Networks expose neighborhood-level blind spots by learning from the relational fabric of the city how infrastructure, income, and proximity co-evolve so planners can prioritize where coverage and capacity will close real gaps rather than paint the map. On top of diagnosis, the chapter operationalizes allocation: fairness-constrained reinforcement learning guides subsidies, training, and infrastructure upgrades in real time, balancing impact with budget while preventing gains from concentrating in already-advantaged areas.

Institutional capacity is treated as a design variable. Digital equity offices are cast as multi-agent systems that coordinate ISPs, libraries, schools, and city departments through shared objectives, stable policy updates, and value decomposition that preserves systemic fairness. Libraries emerge as decentralized literacy engines, using federated and meta-learning to adapt instruction to local needs without exporting sensitive data. Regional disparity is addressed through geographically weighted multi-task learning, aligning national and municipal strategies with on-the-ground heterogeneity. Finally, an infrastructure expansion module combines ant-colony search, deep RL, and graph learning to site broadband nodes where they produce the greatest equity per euro, while Bayesian reinforcement learning steers funding under uncertainty and makes progress auditable over time.

The chapter offers a civic engineering playbook: quantify inequity, reveal it in space, act on it fairly, and institutionalize learning so digital transformation expands capability, dignity, and opportunity for everyone, not just the already connected.

D. Sargiotis, *Empathy in Algorithms*, https://doi.org/10.1007/978-3-032-32128-2_7

7.1 Introduction

The digital divide refers to the gap between individuals who have access to modern information and communication technology (ICT) and those who do not. This gap disproportionately affects marginalized communities, often based on factors such as socioeconomic status, geographic location, race, and age (Feurich et al. 2024). In urban populations, the divide manifests in disparities in access to technology, broadband connectivity, and digital literacy. These barriers hinder the ability of certain groups to fully participate in digital society, exacerbating inequalities in education, employment, and access to services (San Francisco Digital Equity Strategic Plan 2019). Addressing these barriers is critical for fostering inclusive development in smart cities.

To address the digital divide in AI-powered urban environments, we propose a Digital Equity Index Optimization Model. This model quantifies disparities in digital access and literacy, then guides equitable resource allocation to minimize structural gaps in connectivity, infrastructure, and ICT usage. It enables urban AI systems to proactively identify underserved regions and demographics and adjust investment, training, or service delivery accordingly, supporting inclusive digital participation in education, employment, and public services.

At the core of this model is the Digital Equity Index (DEI) for population group gg in region rr:

$$\mathrm{DEI}_{g,r} = w_1 \cdot \mathrm{Access}_{g,r} + w_2 \cdot \mathrm{Connectivity}_{g,r} + w_3 \cdot \mathrm{Literacy}_{g,r}$$

where $\mathrm{Access}_{g,r}$ is the proportion of individuals with access to ICT devices (e.g., smartphones, laptops), $\mathrm{Connectivity}_{g,r}$ r is the availability and quality of broadband infrastructure, $\mathrm{Literacy}_{g,r}$ is a normalized score of digital skills, w_1, w_2, w_3 are policy-defined weights representing the importance of each dimension.

To reduce inequity, an AI-driven targeted allocation policy minimizes the disparity in DEI across all groups and regions:

$$\underset{\alpha}{min} \underset{(g_1,r_1),(g_2,r_2)}{max} |\mathrm{DEI}_{g_1,r_1}(\alpha) - \mathrm{DEI}_{g_2,r_2}(\alpha)|$$

where α represents the intervention strategy (e.g., investment, training programs, hardware provision), and $\mathrm{DEI}_{g,r}(\alpha)$ is the resulting equity score after applying intervention α. The objective ensures a minimax reduction in digital disparity across all social and spatial combinations.

To enforce minimum thresholds for vulnerable groups, we introduce a digital inclusion constraint:

$$\mathrm{DEI}_{g,r}(\alpha) \geq \eta, \forall (g, r) \in V$$

where $\mathcal{V}$ is the set of vulnerable group-region pairs (e.g., elderly in rural zones), and η is the minimum acceptable digital equity score defined by policy.

This AI-driven model enables public sector agencies to systematically identify and close digital access gaps through data-driven targeting and fairness constraints. By decomposing digital inclusion into measurable components and optimizing equity interventions, it ensures that smart city services are not merely efficient but also just and inclusive. The framework ensures that policies for broadband expansion, ICT access, and digital literacy training are aligned with the structural realities of marginalized populations. It moves beyond reactive planning and enables proactive, algorithmically guided governance transforming digital equity from a rhetorical ideal into an operational guarantee.

Framework: Digital Equity Index Optimization Framework (DEIOF)
Layered AI Architecture for Reducing the Digital Divide and Promoting Inclusivity

Layer	Main functionality	AI module(s) and components	Operational scope
Layer 1	Preprocessing data for digital inclusion and equity	Digital inclusion data preprocessing engine	Input → Raw socio-economic, digital literacy, and infrastructure data → Output → Processed data for further analysis
Layer 2	Digital equity index (DEI) calculation for population groups	Digital equity index (DEI) calculation	Input → Demographic and infrastructure data → Output → Digital equity score for each population group
Layer 3	Minimization of digital disparity across regions	Minimax disparity minimization	Input → Digital equity scores, intervention strategies → Output → Optimized resource allocation minimizing disparity
Layer 4	Enforcement of digital inclusion thresholds	Digital inclusion constraint	Input → Equity scores, population data → Output → Ensured minimum thresholds for vulnerable groups
Layer 5	Proactive policy adjustment and real-time feedback	Adaptive equity policy module	Input → Real-time system feedback, performance data → Output → Adaptive policies for equitable digital inclusion

7.2 Investigating the Impact of Socioeconomic Factors

The digital divide, particularly within urban populations, is significantly influenced by socioeconomic factors such as income, education, and geographic location. These elements not only determine the extent of access to technology but also the capacity of individuals and communities to benefit from digital innovations. Based on the following analysis, we can examine how these factors create barriers and deepen the digital divide in urban settings.

7.2.1 *Income and Affordability of Technology*

Income level is one of the most critical determinants of digital access. Lower-income households often lack the financial means to purchase essential technology, such as computers, smartphones, and high-speed internet access. This financial barrier leaves a significant portion of the population disconnected from the digital world. In urban environments, where technological infrastructure may be available, it is still underutilized by those unable to afford devices or the ongoing costs of connectivity (San Francisco Digital Equity Strategic Plan 2019). The San Francisco Digital Equity Strategic Plan highlights how affordability challenges disproportionately affect marginalized communities, leading to a situation where they are excluded from many aspects of digital life, including education, employment opportunities, and access to government services. This exclusion perpetuates socioeconomic disparities, as those without access are further disadvantaged in a rapidly digitizing economy.

Additionally, individuals in low-income neighborhoods are often limited to mobile-only internet access, which lacks the capabilities of broadband services. This restricted access impedes activities such as remote work, online learning, and accessing comprehensive digital resources, which require more robust, stable internet connections (San Francisco Digital Equity Strategic Plan 2019).

To address affordability as a dynamic barrier to equitable digital participation, we introduce a Deep Q-Learning algorithm to optimize subsidy allocation in income-diverse urban regions. This reinforcement learning framework enables the AI system to learn adaptive affordability strategies based on long-term digital inclusion outcomes, optimizing financial resource deployment under budget constraints. It allows the system to balance access maximization with income disparities while responding to changes in urban affordability demands over time.

At the core of this model lies the Q-value function:

$$Q(s_t, a_t) = r_t + \gamma \cdot \underset{a'}{max} Q(s_{t+1}, a')$$

where $Q(s_t, a_t)$ is the expected cumulative reward of taking action a_t in state s_t, r_t is the immediate reward representing affordability-adjusted access utility, γ is the discount factor capturing the importance of future equity, and $\underset{a'}{max} Q(s_{t+1}, a')$ is the best future value from the next state s_{t+1}.

The reward function guiding the learning process is defined as:

$$r_t = \alpha \cdot \text{Access}_t - \beta \cdot \text{Cost}_t$$

where Access_t quantifies the number of low-income users gaining broadband access at time t, Cost_t is the total subsidy expenditure, and α, β are scaling coefficients that prioritize digital inclusion while penalizing inefficiency.

To ensure fairness across income groups, we integrate a parity-adjusted access constraint into the learning environment:

$$\Delta_{income} = |\mathbb{E}_{i \in H}[Q(s, a)] - \mathbb{E}_{i \in L}[Q(s, a)]| \leq \delta$$

where H and L represent high-income and low-income households respectively, and δ is the maximum allowed access utility disparity between these groups to prevent skewed benefit distributions.

This Deep Q-Learning framework enables AI systems to continuously learn subsidy policies that are not only cost-effective but also equity-aligned. By balancing short-term access gains with long-term income-based fairness constraints, the model allows smart city infrastructures to dynamically adapt to affordability gaps, support digital participation in underprivileged communities, and avoid reinforcing socioeconomic inequalities.

Framework: Deep Q-Learning Subsidy Allocation Framework (DQLSAF)
Layered AI Architecture for Affordability and Digital Inclusion in Smart Cities

Layer	Main functionality	AI module(s) and components	Operational scope
Layer 1	Data collection and preprocessing for affordability analysis	Affordability data preprocessing engine	Input → Raw socio-economic and digital access data → Output → Processed data for further analysis
Layer 2	Q-value function calculation for subsidy allocation	Q-value function (deep Q-learning)	Input → Demographic data, subsidy distribution → Output → Q-value function for subsidy actions
Layer 3	Reward function calculation for access and cost optimization	Reward function	Input → Access utility, cost data → Output → Reward calculation for system's subsidy allocation
Layer 4	Fairness enforcement through income parity constraint	Fairness constraint (parity-adjusted access)	Input → Q-values for high and low-income households → Output → Fair access disparity constraint
Layer 5	Real-time adaptation and policy learning	Adaptive policy learning module	Input → Real-time system feedback, performance data → Output → Adaptive policy recommendations for equitable digital access

7.2.2 *Education and Digital Literacy*

Education plays a critical role in determining an individual's ability to navigate and benefit from digital technologies. Those with lower educational attainment are less likely to possess the digital literacy skills necessary to engage with online platforms. The Bridging the Urban–Rural Digital Divide report (Feurich et al. 2024) identifies that educational gaps, especially in underserved communities, correlate with limited digital proficiency. This deficiency hinders individuals from utilizing technology to its fullest potential, further widening the gap between those who can effectively engage with digital tools and those who cannot.

In many cases, public and private initiatives have aimed to offer digital literacy programs to reduce this divide. However, access to these educational resources is often limited in low-income urban areas where such programs are most needed (Feurich et al. 2024). Without sufficient access to these resources, marginalized communities struggle to leverage the benefits of smart city technologies or fully participate in the digital economy.

To systematically address the intersection of educational disparities and digital exclusion in underserved urban populations, we introduce a Deep Reinforcement Learning (DRL) framework optimized for dynamic resource allocation in digital literacy interventions. This model learns optimal strategies for distributing digital education programs to areas of greatest need, considering temporal patterns in access, engagement levels, and infrastructure limitations. At its core, the DRL agent adapts over time through iterative feedback, refining its policies to maximize societal digital engagement under resource constraints. The agent's policy function, updated via the Proximal Policy Optimization (PPO) method, a robust and stable variant of DRL, optimizes long-term cumulative rewards associated with improved digital proficiency across spatial and demographic segments.

At the core of this framework lies the Proximal Policy Optimization objective function, which ensures stable policy updates while maximizing the expected reward over time:

$$L^{CLIP}(\theta) = \mathbb{E}_t[min(r_t(\theta)\hat{A}_t, \text{clip}(r_t(\theta), 1-\epsilon, 1+\epsilon)\hat{A}_t)]$$

where $L^{CLIP}(\theta)$ is the clipped surrogate objective that the policy network seeks to maximize, $r_t(\theta) = \frac{\pi_\theta(a_t|s_t)}{\pi_{\theta_{\text{old}}}(a_t|s_t)}$ is the probability ratio between the current and previous policy, $\hat{A}_t$ is the advantage estimate at timestep t, ϵ is the clipping parameter controlling how far the new policy can deviate, and clip $(\cdot)$ limits the objective to prevent large policy updates that may destabilize learning.

This function allows the system to update its policy network safely, ensuring that the model does not overreact to noisy advantage estimates while still improving decisions that increase digital engagement outcomes in underserved areas. The advantage term $\hat{A}_t$ quantifies the relative quality of the chosen action over the average expected return, guiding the agent to favor strategies such as targeting specific schools or neighborhoods that demonstrate higher educational impact.

To estimate the advantage term used above, we employ the Generalized Advantage Estimation (GAE), which balances bias and variance for more stable policy learning:

$$\hat{A}_t = \sum_{l=0}^{T-t} (\gamma\lambda)^l \delta_{t+l}$$

where $\hat{A}_t$ is the estimated advantage at timestep t, γ is the discount factor for future rewards, λ is the smoothing parameter controlling the bias-variance trade-off, δ_{t+l} is the temporal difference error defined as $r_{t+l} + \gamma V(s_{t+l+1}) - V(s_{t+l})$, and T is the horizon of the episode. This formulation accumulates the temporal difference errors over time, emphasizing more recent observations while incorporating long-term feedback. It ensures that the model emphasizes impactful interventions such as timely mobile-based digital training while de-emphasizing short-term anomalies or one-off gains.

The system's overall expected return under the learned policy is then evaluated through the cumulative reward function:

$$J(\theta) = \mathbb{E}_{\pi_\theta}[\sum_{t=0}^{T} \gamma^t r_t]$$

where $J(\theta)$ is the expected total reward under policy π_θ, γ is the discount factor, r_t is the reward obtained at time t, and T is the length of the episode during which digital literacy programs are deployed and evaluated. This cumulative reward guides the long-term planning behavior of the AI system, enabling it to allocate digital education programs in a manner that maximizes digital participation and minimizes systemic exclusion over extended time horizons.

This AI-driven framework learns and adapts optimal policies for targeting digital education initiatives. It enables city planners and policymakers to prioritize regions and populations where educational intervention will yield the highest digital inclusivity gains. As a result, it transforms passive data on educational disparity into actionable policy that narrows the digital divide and empowers marginalized urban populations to participate fully in smart city ecosystems.

Framework: Deep Reinforcement Learning (DRL) for Digital Literacy Resource Allocation (DRLR)

Layered AI Architecture for Dynamic Resource Allocation and Digital Inclusion

Layer	Main functionality	AI module(s) and components	Operational scope
Layer 1	Data collection and preprocessing for digital literacy programs	Digital literacy data ingestion pipeline	Input → Raw data on digital access, literacy, and socio-economic factors → Output → Preprocessed data ready for analysis
Layer 2	Optimization of digital literacy interventions using PPO	Proximal Policy Optimization (PPO)	Input → Intervention strategies, digital access data → Output → Optimized resource allocation policies
Layer 3	Advantage estimation for policy updates	Generalized advantage estimation (GAE)	Input → Temporal difference errors, rewards → Output → Estimated advantages for stable policy learning
Layer 4	Maximization of long-term rewards for digital inclusion	Cumulative reward function	Input → Cumulative reward data, policy output → Output → Policy updates that maximize long-term digital inclusion

(continued)

(continued)

Layer	Main functionality	AI module(s) and components	Operational scope
Layer 5	Post-AI evaluation and feedback integration	Continuous evaluation and feedback orchestrator	Input → Real-time system feedback, performance data → Output → Iterative improvement and policy adjustments

7.2.3 Geographical Disparities and Infrastructure

Geographic location within urban areas also plays a significant role in the digital divide. While urban settings typically have more extensive digital infrastructure, not all neighborhoods are equally served. The San Francisco Digital Equity Strategic Plan (2019) notes that certain low-income neighborhoods lack sufficient infrastructure for high-speed internet, limiting residents' ability to participate in the digital world.

Moreover, urban–rural distinctions exist even within city boundaries, where certain marginalized urban populations experience conditions more akin to rural digital deprivation due to poor infrastructure development. This geographic inequality exacerbates the digital divide, particularly in neighborhoods where residents are already facing multiple socioeconomic challenges (Feurich et al. 2024).

To address spatial inequities in digital infrastructure across urban neighborhoods, we propose a Graph Neural Network (GNN) framework that models the relational topology of digital infrastructure availability within cities. The GNN effectively captures geographic disparities by representing each neighborhood as a node and their infrastructural interdependencies as edges in a graph structure. Unlike classical models that treat each region in isolation, this graph-based learning approach allows the system to infer latent digital deprivation patterns based on topological proximity, shared socioeconomic features, and connectivity constraints. By learning from both node attributes (e.g., infrastructure score, income level, population density) and graph structure (e.g., adjacency to well-connected areas), the GNN enables policymakers to pinpoint underserved digital zones and prioritize targeted interventions.

At the foundation of this framework lies the neighborhood aggregation function that defines how each node updates its digital infrastructure representation over time:

$$h_v^{(k+1)} = \sigma\Big(\sum_{u \in \mathcal{N}(v)} W^{(k)} h_u^{(k)} + b^{(k)} \Big)$$

where $h_v^{(k+1)}$ is the updated embedding of node v at iteration $k + 1$, $\mathcal{N}(v)$ represents the set of neighboring nodes connected to v, $h_u^{(k)}$ is the feature vector of neighbor u at layer k, $W^{(k)}$ is the learnable weight matrix applied during aggregation, $b^{(k)}$ is the bias term, and σ is a non-linear activation function (e.g., ReLU) that enables learning of complex interactions. This function allows each neighborhood to update its infrastructure profile by incorporating information from adjacent neighborhoods. As a result, the model learns how geographic proximity to better-connected or similarly underserved areas impacts a region's digital deprivation risk. For instance, a low-income zone adjacent to high-speed connected districts may be more likely to benefit from shared infrastructure spillover, while isolated areas may remain underserved despite proximity to central nodes.

To ensure that the GNN captures both local and global spatial dependencies, the final node-level digital deprivation prediction is generated after multiple propagation layers and is expressed through a softmax classifier:

$$\hat{y}_v = \mathrm{softmax}(W^{(L)} h_v^{(L)} + b^{(L)})$$

where $\hat{y}_v$ is the predicted probability distribution of digital deprivation categories for node v, $h_v^{(L)}$ is the final hidden representation after L GNN layers, $W^{(L)}$ and $b^{(L)}$ are the learned weight matrix and bias for the classification layer, and softmax $(\cdot)$ converts the outputs into normalized probabilities across defined classes (e.g., high, medium, low digital access).

This final function assigns each neighborhood a digital accessibility score based on the learned graph embeddings, revealing zones of compounded infrastructural neglect. These insights empower city governments to prioritize specific neighborhoods for broadband expansion, targeted subsidies, or community-based Wi-Fi initiatives.

This AI-driven framework provides a principled method for identifying infrastructural blind spots within cities. It enables public agencies to move beyond static, map-based assessments toward predictive, data-driven policy actions. The graph-based structure captures urban heterogeneity with respect to infrastructure diffusion and spatial connectivity, ensuring that digital equity planning becomes both geographically precise and socially just.

Framework: Graph Neural Network (GNN) for Digital Infrastructure Equity (GNNDIE)
Layered AI Architecture for Identifying Geographic Disparities in Digital Access

Layer	Main functionality	AI module(s) and components	Operational scope
Layer 1	Data collection and preprocessing for digital infrastructure disparity	Digital Infrastructure Data Preprocessing Engine	Input → Raw data on digital infrastructure, population density, and income levels → Output → Processed data for analysis
Layer 2	Neighborhood aggregation function for digital infrastructure	Graph Neural Network (GNN)	Input → Neighborhoods, infrastructure data → Output → Updated digital infrastructure representation for each node
Layer 3	Final node-level digital deprivation prediction using softmax	Softmax classifier	Input → GNN node representations → Output → Predicted probability distribution of digital deprivation categories for neighborhoods

(continued)

(continued)

Layer	Main functionality	AI module(s) and components	Operational scope
Layer 4	Geographic disparity minimization and resource allocation	Geographic equity allocation module	Input → Node-level predictions, adjacency data → Output → Resource allocation to underserved areas based on GNN insights
Layer 5	Feedback and iterative improvement	Adaptive policy feedback engine	Input → Real-time system feedback, resource allocation outcomes → Output → Iterative improvements in policy recommendations

7.2.4 Employment and Economic Mobility

Access to technology also has a direct impact on employment opportunities. The digital divide limits the ability of individuals to participate in remote work, upskill through online courses, or access job opportunities that are increasingly posted and managed through online platforms. Low-income individuals without access to technology are not only excluded from these opportunities but also face greater challenges in economic mobility, perpetuating cycles of poverty (Feurich et al. 2024).

The employment sector is increasingly reliant on digital tools, and those who lack basic digital literacy are at a disadvantage in the labor market. Feurich et al. (2024) emphasize that job-seeking behaviors have shifted online, meaning individuals without regular access to the internet are often unable to compete for jobs, attend virtual interviews, or complete online applications.

To systematically address the digital divide's influence on employment accessibility and upward economic mobility, we introduce a Deep Q-Network (DQN) framework that dynamically learns optimal pathways to digital job accessibility based on an individual's resource constraints, digital literacy level, and geographic disadvantage. As online employment processes increasingly require timely digital interactions, this reinforcement learning-based model enables policymakers and employment agencies to recommend context-specific interventions such as mobile device provisioning, broadband subsidies, or targeted digital training that maximize job accessibility outcomes over time.

The DQN agent learns a state-action value function (Q-function) that evaluates the expected cumulative benefit of choosing a particular support strategy for an individual in a given socioeconomic state. It does so while accounting for delayed rewards, such as eventual job attainment or skill certification, that result from earlier digital support interventions.

At the heart of this framework lies the Bellman update equation, adapted for Deep Q-Networks, which governs the agent's learning process through temporal difference optimization:

$$Q(s_t, a_t) \leftarrow Q(s_t, a_t) + \alpha[r_t + \gamma \underset{a'}{max} Q(s_{t+1}, a') - Q(s_t, a_t)]$$

where $Q(s_t, a_t)$ is the estimated value of taking action a_t in state s_t, α is the learning rate controlling update magnitude, r_t is the immediate reward received after taking action a_t, γ is the discount factor weighing future rewards, and $\underset{a'}{max} Q(s_{t+1}, a')$ is the maximum predicted value for the next state s_{t+1} across all possible actions a'. This equation allows the DQN to learn which intervention strategies (e.g., digital device grants, job portal training) yield the highest long-term employment benefits for individuals constrained by limited technology access. It balances immediate digital support benefits with delayed socioeconomic gains, ensuring that the AI agent does not prioritize short-term fixes over long-term impact.

The DQN approximates the Q-function using a deep neural network parameterized by θ, updated through backpropagation to minimize the loss between predicted and target Q-values:

$$L(\theta) = \mathbb{E}_{(s_t, a_t, r_t, s_{t+1})}[(y_t^{DQN} - Q(s_t, a_t; \theta))^2]$$

where $\mathcal{L}(\theta)$ is the loss function for training the Q-network, $Q(s_t, a_t; \theta)$ is the network-predicted Q-value for the current state-action pair, and $y_t^{DQN} = r_t + \gamma \underset{a'}{max} Q(s_{t+1}, a'; \theta^-)$ is the target Q-value computed using a separate target network θ^- to stabilize learning. This loss function ensures that the Q-network accurately reflects the value of different support actions under varying employment contexts. By minimizing prediction error over time, the model becomes increasingly precise in recommending digital inclusion strategies that lead to job acquisition, digital credentialing, or participation in virtual employment ecosystems.

This AI-driven model enables a transition from static policy interventions to personalized, data-informed actions that optimize long-term economic inclusion. The model effectively navigates the complex, delayed reward structure inherent in employment mobility, ensuring that individuals facing digital exclusion are guided along the most beneficial support trajectories toward sustained labor market integration. This framework provides city governments, workforce agencies, and NGOs with a powerful AI toolkit to close digital gaps not only in access, but in economic opportunity.

Framework: Deep Q-Network (DQN) for Digital Job Accessibility (DQN-DJA)

Layered AI Architecture for Addressing the Digital Divide in Employment Opportunities

Layer	Main functionality	AI module(s) and components	Operational scope
Layer 1	Data Collection and Pre-Processing for Digital Job Accessibility	Digital Employment Data Preprocessing Engine	Input → Raw data on digital access, literacy, income levels, and geographic location → Output → Preprocessed data ready for analysis

(continued)

(continued)

Layer	Main functionality	AI module(s) and components	Operational scope
Layer 2	Optimization of Digital Literacy Interventions Using PPO	Deep Q-Network (DQN)	Input → Intervention strategies, digital access data → Output → Optimized resource allocation policies for digital inclusion
Layer 3	Maximization of Long-Term Rewards for Digital Inclusion	Cumulative Reward Function	Input → Cumulative reward data, policy output → Output → Policy updates that maximize long-term digital engagement outcomes
Layer 4	Feedback and Iterative Improvement	Adaptive Policy Refinement Module	Input → Real-time system feedback, resource allocation outcomes → Output → Iterative improvements in policy recommendations

Figure 7.1 outlines how socioeconomic factors shape the digital divide, focusing on four main areas: income and affordability of technology, education and digital literacy, geographical disparities and infrastructure, and employment and economic mobility. Income and affordability are presented as key barriers, as lower-income households struggle to afford essential technology, mobile-only internet restricts remote work and online learning, and these limitations deepen socioeconomic disparities. Education and digital literacy are also central, since lower educational attainment correlates with limited digital proficiency, while restricted access to digital literacy programs in low-income urban areas hinders engagement with digital tools and smart technologies. Geographical disparities and infrastructure show that some urban neighborhoods lack robust digital infrastructure, face unequal access to high-speed internet, and experience forms of digital deprivation similar to rural areas. Employment and economic mobility are affected because limited digital access restricts job-seeking and remote work, reduces competitiveness in the digital labor market, and perpetuates poverty cycles and limited mobility. Overall, the figure presents the digital divide as a multidimensional socioeconomic problem shaped by affordability, education, infrastructure, and labor-market access.

Socioeconomic factors such as income, education, and geographic location significantly impact access to technology in urban populations, perpetuating the digital divide. As cities continue to implement smart infrastructure, it is vital to address these barriers through targeted interventions that enhance affordability, improve digital literacy, and ensure equitable infrastructure development. Without such measures, the digital divide will continue to exacerbate existing socioeconomic inequalities.

7.3 Initiatives Aimed at Bridging the Digital Divide

To bridge the digital divide, various initiatives have been launched to enhance access to technology and digital literacy.

Fig. 7.1 Impact of socioeconomic factors on the digital divide

7.3.1 Public–Private Partnerships

A key strategy employed to reduce the digital divide is the formation of public–private partnerships. Cities have recognized that the collaboration between public authorities and private sector companies, particularly internet service providers (ISPs), is vital to achieving universal access to affordable, high-speed internet. The San Francisco Digital Equity Strategic Plan (2019) exemplifies this approach by detailing how the city worked with ISPs to extend broadband services to low-income households at affordable rates. This initiative involves offering subsidized internet services and free digital devices to economically disadvantaged residents, ensuring that they can connect to essential online resources.

The strategic plan also mentions partnerships with technology companies and philanthropic organizations to secure donated devices for underserved communities, enabling access to the tools needed for educational and employment opportunities. This ensures that the most vulnerable populations are not left behind in the digital age (San Francisco Digital Equity Strategic Plan 2019).

7.3.2 Community-Based Approaches

Community-based approaches are central to bridging the digital divide. According to the San Francisco Digital Equity Strategic Plan, successful digital inclusion efforts often require engaging directly with affected communities to understand their specific needs and barriers to technology access. The city engaged in community-led initiatives, which involved surveys and community fairs, allowing residents to voice their concerns and contribute to the development of tailored solutions.

Such initiatives go beyond merely providing access to technology they focus on equipping individuals with the digital skills necessary to use technology effectively. Digital literacy programs, for example, are designed to improve technical proficiency among marginalized groups, such as low-income families, seniors, and people with disabilities. These programs teach participants how to use the internet, apply for jobs online, and engage with digital government services (San Francisco Digital Equity Strategic Plan 2019).

To support scalable yet privacy-preserving digital literacy interventions across diverse urban communities, we propose a Federated Meta-Learning (FedMeta) framework. This model enables localized AI systems to collaboratively learn optimal digital training strategies from decentralized community data without ever sharing raw data while adapting rapidly to the distinct needs of marginalized user groups. Each community node (e.g., a neighborhood center or local NGO) trains a local model on its specific digital skill usage data and contributes only model parameters to a central meta-learner, which in turn refines and disseminates an adaptive global strategy for personalized learning interventions.

At the heart of this framework lies the Model-Agnostic Meta-Learning (MAML) update function, which equips the global model with the ability to generalize across community-specific tasks and quickly adapt to new digital learning environments:

$$\theta' = \theta - \alpha \nabla_\theta \mathcal{L}_{\text{local}}(f_\theta)$$

where θ is the global model's initial parameter set, θ' is the adapted local model after a single gradient descent step, α is the adaptation learning rate, $\mathcal{L}_{\text{local}}(f_\theta)$ is the loss incurred by the model f_θ on local digital literacy data (e.g., tutorial completions, help request frequencies), and ∇_θ denotes the gradient with respect to the model parameters.

This function allows each community node to customize the global model to its own user base using minimal data, ensuring rapid adaptation to the needs of, for instance, digitally underserved senior populations in one area and unemployed youth in another.

Once all communities have locally updated their models, the central server aggregates these updates to refine the global model through the Federated Averaging function:

$$\theta \leftarrow \sum_{k=1}^{K} \frac{n_k}{n} \theta'_k$$

where θ is the updated global parameter vector, θ'_k is the locally adapted model from community node k, nknk is the number of training samples at node k, $n = \sum_{k=1}^{K} n_k$ is the total number of training samples across all K participating communities.

This aggregation ensures that each community contributes proportionally to the global knowledge base according to the scale of its training data. It allows the global model to evolve in a way that reflects the diversity of urban digital literacy needs without compromising user privacy or centralizing sensitive data.

The effectiveness of each local model is measured via a meta-objective function, which evaluates the global model's performance across all community tasks:

$$\min_{\theta} \sum_{k=1}^{K} \mathcal{L}_{\text{local}_k}(f_{\theta'_k})$$

where $\mathcal{L}_{\text{local}_k}$ represents the loss on the post-adaptation model $f_{\theta'_k}$ for community k, capturing how well the global model can adapt to each local context after one or a few updates.

This objective ensures that the global model is not merely averaging updates but optimizing for cross-community generalizability and rapid local personalization. It rewards model parameters that minimize the learning burden on individual communities, particularly those with fewer resources or limited data.

This AI-driven model provides an ethical framework for scaling digital literacy programs across heterogeneous urban communities. It respects data sovereignty and local autonomy while continuously learning how to serve evolving community needs. In real-world deployments, this system enables cities to deliver tailored digital skill-building interventions in neighborhoods ranging from dense immigrant districts to isolated senior communities ensuring no group is algorithmically neglected in the push toward digital equity.

Framework: Federated Meta-Learning Framework for Digital Literacy (FedMeta-DL)
Layered AI Architecture for Privacy-Preserving Digital Literacy Interventions

Layer	Main functionality	AI module(s) and components	Operational scope
Layer 1	Data Collection: Gather local digital literacy data	Local Data Preprocessing Engine	Input → Community-specific digital literacy data (e.g., tutorial completions) → Output → Local model updates
Layer 2	Learning: Local training of models with community data	Model-Agnostic Meta-Learning (MAML)	Input → Digital literacy data → Output → Locally trained model with updated parameters
Layer 3	Adaptation: Model adaptation to community-specific tasks	Model-Agnostic Meta-Learning (MAML)	Input → Local community model → Output → Adapted model for local needs

(continued)

(continued)

Layer	Main functionality	AI module(s) and components	Operational scope
Layer 4	Aggregation: Federated Averaging to refine the global model	Federated Averaging (FedAvg)	Input → Local model parameters → Output → Global model refinement
Layer 5	Meta-Objective: Evaluation of the global model's performance	Global Model Evaluation Module	Input → Global model → Output → Performance evaluation and cross-community optimization
Layer 6	Deployment: Implementation of the global model for personalized interventions	Personalized Deployment Orchestrator	Input → Global model → Output → Tailored digital literacy interventions across communities
Layer 7	Monitoring: Ongoing evaluation and model refinement based on feedback	Adaptive Federated Monitoring Engine	Input → Feedback from community nodes → Output → Updated global model and adaptive interventions

7.3.3 *Infrastructure Expansion and Digital Inclusion Networks*

Another important initiative aimed at bridging the digital divide is infrastructure expansion, particularly in underserved areas. The Feurich et al. (2024) report discusses how cities and regions have made significant investments in upgrading and expanding their digital infrastructures. For example, European Union initiatives, particularly the Rural Vision 2040, focus on improving broadband networks in areas that are traditionally under-resourced. This includes not only rural regions but also low-income urban neighborhoods where digital infrastructure may be lacking or outdated.

Furthermore, cities like San Francisco have developed Digital Inclusion Networks, which bring together stakeholders from across public, private, and nonprofit sectors to collaborate on digital equity initiatives. These networks are designed to identify gaps in digital infrastructure and create coordinated plans to address them, focusing on expanding Wi-Fi access points and providing residents with free or low-cost internet options (San Francisco Digital Equity Strategic Plan 2019).

To address this challenge, we propose a state-of-the-art AI-driven framework that employs Deep Reinforcement Learning (DRL) and Graph Neural Networks (GNNs), alongside Ant Colony Optimization (ACO), to optimize the deployment of digital infrastructure. Unlike traditional metaheuristics, the integration of DRL provides an adaptive learning mechanism that can continuously adjust based on real-time feedback, while GNNs are introduced to capture the spatial dependencies between network nodes, enhancing the system's understanding of urban geography. This AI-driven approach, utilizing the synergy between ACO, DRL, and GNNs, ensures that infrastructure expansion is not only efficient but also maximizes coverage in underserved areas while minimizing cost and deployment time. The Ant Colony Optimization (ACO) algorithm is traditionally used to identify optimal paths in a network. In our framework, ACO will guide artificial agents (ants) to identify the most efficient deployment paths for broadband nodes. The probability that an agent chooses a particular path from node j to node i at any given iteration is influenced by both the pheromone level $\tau_{ij}(t)$ and the heuristic desirability $\tau_{ij}(t)$, which reflects the cost or distance associated with the path.

The Transition Rule for path selection can be expressed mathematically as:

$$P_{ij}(t) = \frac{[\tau_{ij}(t)]^{\alpha}[\eta_{ij}]^{\beta}}{\sum_{k\in\mathcal{N}_j}[\tau_{ik}(t)]^{\alpha}[\eta_{ik}]^{\beta}}$$

where $P_{ij}(t)$ is the probability of an ant moving from node j to node i at time t, $\tau_{ij}(t)$ is the pheromone level on edge (i, j), η_{ij} is the heuristic desirability of that edge (typically inversely proportional to cost or distance), α controls the influence of pheromone trails, β controls the influence of heuristic information, and $\mathcal{N}_j$ is the set of neighboring candidate nodes from node j. This function models how the AI agent probabilistically chooses where to expand infrastructure based on both historical reinforcement and local feasibility.

To improve the framework's capability to understand the spatial layout of urban areas, we integrate Graph Neural Networks (GNNs). GNNs allow us to model the spatial dependencies between network nodes, enhancing the system's ability to optimize node placements by taking into account both local and global spatial features. This is critical for ensuring that infrastructure deployment is efficient and aware of geographical constraints, such as barriers or unconnected zones.

The GNN model's Node Representation Update can be formulated as:

$$h_i^{(l+1)} = \sigma(W^{(l)} \cdot (h_i^{(l)} + \sum_{j\in N(i)} A_{ij}h_j^{(l)}))$$

where $h_i^{(l+1)}$, updated node representation for node i at layer $l+1$, $h_i^{(l)}$, node representation for node i at layer l, $W^{(l)}$, weight matrix for layer l, $N(i)$, set of neighboring nodes of node i, A_{ij}, adjacency matrix element between nodes i and j, σ, activation function (typically ReLU). This function captures the influence of neighboring nodes on the current node's representation, enhancing the model's understanding of the spatial relationship between nodes, thus improving decision-making for infrastructure placement.

The final enhancement in our AI-driven framework is Deep Reinforcement Learning (DRL), which allows the model to continuously learn optimal deployment strategies through interaction with the environment. Using the Proximal Policy

Optimization (PPO) algorithm, agents can efficiently navigate the space of possible deployment paths, optimizing for cost, coverage, and equity in real time.

The PPO Objective Function for policy optimization is expressed as:

$$L(\theta) = \mathbb{E}_t[min(r_t(\theta)\hat{A}_t, \text{clip}(r_t(\theta), 1 - \epsilon, 1 + \epsilon)\hat{A}_t)]$$

where $L(\theta)$, objective function to be maximized, $r_t(\theta)$, probability ratio of the current policy to the old policy, $\hat{A}_t$, advantage function at time t, ϵ, clipping parameter (controls the magnitude of policy updates), $\mathbb{E}_t$, expectation over time steps. This function allows the model to adjust its policy based on feedback, improving the efficiency of the infrastructure deployment strategy and adapting to new environmental conditions.

As artificial agents iteratively evaluate paths through the urban digital grid, the pheromone trails are updated to reinforce high-quality paths that provide better coverage to underserved areas. The pheromone update rule, which governs learning in the system, is given by:

$$\tau_{ij}(t+1) = (1-\rho) \cdot \tau_{ij}(t) + \sum_{k=1}^{m} \Delta\tau_{ij}^{k}(t)$$

where $\tau_{ij}(t+1)$ is the updated pheromone on edge (i, j), $\rho \in [0,1]$ is the evaporation coefficient controlling memory decay, $\Delta\tau_{ij}^{k}(t)$ is the amount of pheromone deposited by ant k on that edge at time t, and m is the total number of ants. This function enables the system to learn optimal deployment paths by reinforcing successful strategies while allowing for exploration and adaptation to emerging infrastructure gaps.

This AI-driven framework, integrating ACO, GNN, and DRL, provides a solution to the challenge of digital infrastructure expansion in underserved regions. The ACO component optimizes path selection, the GNN module captures spatial dependencies, and DRL enables real-time adaptation and learning of optimal deployment strategies. The careful integration of these state-of-the-art AI models ensures that the infrastructure expansion is not only efficient but also maximally equitable, addressing cost, coverage, and accessibility concerns in a holistic manner.

Framework: AI-Driven Digital Infrastructure Optimization Framework (AIDIOF)
Layered AI Architecture for Optimizing Broadband Expansion

Layer	Main functionality	AI module(s) and components	Operational scope (Input → Processing → Output)
Layer 1	Data Collection: Gathering input data (e.g., spatial grid, network topology)	Infrastructure Data Preprocessing Engine	Input → Spatial grid, network topology → Output → Preprocessed input for AI models

(continued)

(continued)

Layer	Main functionality	AI module(s) and components	Operational scope (Input → Processing → Output)
Layer 2	Path Selection: Optimization using Ant Colony Optimization (ACO) for efficient infrastructure expansion	Ant Colony Optimization (ACO)	Input → Spatial grid, network topology → Output → Optimized path selections for infrastructure
Layer 3	Learning: Graph Neural Networks (GNN) for spatial understanding and network structure optimization	Graph Neural Networks (GNN)	Input → Network structure, spatial dependencies → Output → Node placement optimization
Layer 4	Optimization: DRL Policy optimization for infrastructure deployment decision-making	Deep Reinforcement Learning (DRL)	Input → Current deployment state, feedback → Output → Optimized deployment decisions
Layer 5	Pheromone Update: Ant Colony Optimization (ACO) reinforcement of successful paths	Ant Colony Optimization (ACO)	Input → Path information, feedback → Output → Reinforced paths for future iterations
Layer 6	Monitoring: Ongoing evaluation of system performance and infrastructure coverage	Network Performance Monitoring Module	Input → Performance data, failure detection → Output → Monitoring reports and system evaluation

7.3.4 *Affordable Internet and Device Distribution*

Several programs have focused on making internet access more affordable and providing digital devices to those who cannot afford them. According to Feurich et al. (2024), device donation programs have been highly effective in addressing the equipment aspect of the digital divide. These programs collect surplus devices from businesses and refurbish them for distribution to low-income households and schools. Ensuring that people have the necessary hardware, such as laptops and tablets, is essential to closing the digital divide.

Additionally, low-cost broadband programs as detailed in both Feurich et al. (2024) and the San Francisco Digital Equity Strategic Plan (2019) are pivotal. These programs typically involve negotiations between city governments and ISPs to provide discounted internet service plans to low-income households. The goal is to ensure that every resident has access to reliable, affordable internet regardless of their economic status.

To support equitable device and broadband distribution under complex logistical and economic constraints, we introduce a Multi-Objective Genetic Algorithm (MOGA) framework. MOGA enables the optimization of parallel objectives such as maximizing the number of low-income households reached, minimizing device delivery times, and minimizing ISP service costs by evolving a population of candidate distribution strategies through biologically inspired operators.

Each candidate strategy x_i is defined as a vector of decisions, such as which households receive which refurbished devices, what delivery routes are used, and which broadband plan is offered to each household. The optimization process evaluates these decisions using a multi-objective function:

$$f(x) = (max \sum_{j=1}^{N} u_j(x), min \sum_{k=1}^{M} c_k(x), min \sum_{l=1}^{L} d_l(x))$$

where $u_j(x)$ is a utility score measuring digital access improvement for household j, $c_k(x)$ is the cost associated with broadband plan k, and $d_l(x)$ is the delivery time or logistical burden for device l. The function jointly maximizes inclusion utility while minimizing both economic and logistical costs.

At each generation, candidate solutions are assessed using a multi-objective fitness function that quantifies trade-offs between societal benefit and operational feasibility. The selection probability of each candidate strategy x_i is determined through a Pareto dominance-based ranking scheme:

$$P(x_i) = \frac{1}{R(x_i)+1}$$

where $P(x_i)$ is the probability of selecting solution x_i for crossover and mutation in the next generation, $R(x_i)$ is the Pareto rank of solution x_i, indicating how many other solutions dominate it in terms of both affordability and reach. A lower rank reflects greater fitness, prioritizing strategies that offer broader access and lower cost.

This formulation allows decision-makers to identify and implement efficient, equity-oriented internet and device distribution plans. The genetic algorithm adapts dynamically to real-world constraints such as varying broadband availability, device inventory limits, and household needs ensuring that no single optimization goal dominates at the expense of others. By balancing competing priorities, this AI framework helps cities deliver affordable connectivity and digital access where they are needed most, thus operationalizing fairness within infrastructural limitations.

Framework: Multi-Objective Genetic Algorithm for Digital Divide Optimization (MOGA-DDO)

Layered AI Architecture for Optimized Device and Broadband Distribution

Layer	Main functionality	AI module(s) and components	Operational scope
Layer 1	Initial Setup: Define socio-political constraints and community requirements for device and broadband distribution	Constraint Definition Engine	Input → Socio-political data, community needs → Output → Defined constraints for optimization
Layer 2	Optimization: Multi-objective optimization of resource allocation and distribution strategy using MOGA	Multi-Objective Genetic Algorithm (MOGA)	Input → Device inventory, broadband availability, household data → Output → Optimized resource distribution strategies
Layer 3	Solution Selection: Ranking solutions using Pareto dominance for final selection	Pareto Solution Ranking Module	Input → Candidate solutions, Pareto dominance → Output → Ranked candidate solutions
Layer 4	Evaluation: Continuous assessment and feedback to ensure system effectiveness and alignment with societal goals	Adaptive Optimization Feedback Engine	Input → Evaluation metrics, feedback data → Output → Refined optimization strategies

7.3.5 Education and Digital Literacy Training

The lack of digital literacy is one of the most significant barriers to achieving digital equity. Therefore, cities have implemented education and digital literacy programs aimed at equipping residents with the necessary skills to engage with digital technologies meaningfully. These initiatives target groups such as seniors, people with disabilities, and economically disadvantaged individuals who may have had limited exposure to technology.

The San Francisco Digital Equity Strategic Plan highlights that digital literacy training is often conducted in collaboration with libraries, schools, and community centers, where participants can learn how to navigate the internet, use email, apply for jobs, and access essential online services. These programs are crucial for ensuring that all residents can fully participate in the digital economy (San Francisco Digital Equity Strategic Plan 2019).

To ensure that digital literacy training programs adapt to each learner's pace, background, and environment, we introduce a Reinforcement Learning (RL) framework based on Deep Q-Learning. This AI-driven model functions as an intelligent tutor that learns to recommend the most effective educational activity for each individual at each stage of their learning trajectory. It balances the short-term goal of increasing immediate engagement with the long-term objective of sustained skill acquisition. The core of this adaptive learning process lies in the Q-function update rule, which defines how the model learns from past actions to improve future training decisions:

$$Q(s_t, a_t) \leftarrow Q(s_t, a_t) + \alpha[r_t + \gamma \underset{a'}{max} Q(s_{t+1}, a') - Q(s_t, a_t)]$$

where $Q(s_t, a_t)$ is the estimated value of choosing learning activity a_t in state s_t, which represents the learner's current knowledge and confidence, α is the learning rate that controls how much new feedback influences existing knowledge, r_t is the reward received from the learner's progress or engagement after activity a_t, γ is the discount factor prioritizing future learning gains, and $\underset{a'}{max} Q(s_{t+1}, a')$ is the predicted optimal value of the next learning action a' in the updated learner state s_{t+1}. This function allows the system to continuously update its recommendations for exercises, tutorials, or assessments based on observed learner behavior, such as quiz results, response time, or help requests. In doing so, it enables the system to personalize learning sequences in real time for populations with limited prior exposure to digital tools.

To approximate this decision-making function over a complex, high-dimensional learner profile space, the Deep Q-Network (DQN) is trained via a supervised loss function that minimizes the difference between predicted and target Q-values:

$$L(\theta) = \mathbb{E}_{(s_t, a_t, r_t, s_{t+1})}[(y_t - Q(s_t, a_t; \theta))^2]$$

where $\mathcal{L}(\theta)$ is the loss function used to train the DQN with parameters θ, $Q(s_t, a_t; \theta)$ is the network's current prediction for the state-action pair, and $y_t = r_t + \gamma \underset{a'}{max} Q(s_{t+1}, a'; \theta^-)$ is the target value computed using a stable target network with parameters θ^-. This formulation ensures stable and efficient learning even as user behavior varies across different demographic or ability-based groups.

This ensures that participants whether they are elderly first-time users or young adults with inconsistent schooling receive educational activities that are neither too basic nor too complex. As a result, the system accelerates confidence-building and retention of digital skills, while minimizing frustration or dropout. The AI-driven model supports community-based trainers and facilitators by dynamically adapting lesson sequences, optimizing resource allocation, and flagging learners who may benefit from additional support. This framework transforms static digital literacy programs into living, responsive learning ecosystems that evolve with community needs and individual learning journeys.

Framework: Deep Q-Learning for Adaptive Digital Literacy Training (DQL-ADLT)
Layered AI Architecture for Personalized Digital Literacy Programs

Layer	Main functionality	AI module(s) and components	Operational scope
Layer 1	Setup: Define learner profiles, digital literacy goals, and demographic factors	Learner Profiling and Goal Initialization Engine	Input → Learner data (age, prior knowledge, disabilities) → Output → Defined learner profiles and goals
Layer 2	Learning: Adaptive recommendations for digital literacy activities using Q-learning	Deep Q-Learning (DQL)	Input → Learner profiles, activity options, feedback → Output → Adaptive learning activities recommendations
Layer 3	Training: Loss function to optimize DQN for stable learning and accurate predictions	DQN Training Loss Optimization Module	Input → Learner feedback, Q-function predictions → Output → Optimized learning model parameters
Layer 4	Monitoring: Continuous tracking of learner progress and adjustment of training plans	Progress Monitoring and Plan Adaptation Engine	Input → Learner progress data, feedback → Output → Refined training plans and next activity recommendations

Figure 7.2 presents initiatives aimed at bridging the digital divide through five interconnected areas: public–private partnerships, community-based approaches, infrastructure expansion and digital inclusion networks, affordable internet and device distribution, and education and digital literacy training. Public–private partnerships include collaboration with ISPs to provide affordable high-speed internet, subsidized internet services and device donations for low-income households, and partnerships with technology companies for device donations. Community-based approaches emphasize direct engagement with communities to identify needs, digital literacy programs for marginalized groups such as seniors and low-income families, and community events and surveys to tailor digital inclusion solutions. Infrastructure expansion and digital inclusion networks focus on broadband investments in underserved areas, expanded Wi-Fi access points and low-cost internet options, and cross-sector coordination through digital inclusion networks. Affordable internet and device distribution include low-cost broadband programs, refurbished or donated devices for low-income households and schools, and affordable device distribution to support digital participation. Education and digital literacy training provide skills for navigating the internet, job applications, and online services through partnerships with libraries, schools, and community centers, with targeted programs for seniors and disadvantaged groups. The figure presents digital inclusion as a coordinated effort involving access, infrastructure, affordability, community engagement, and digital skills development.

Initiatives aimed at bridging the digital divide are multifaceted and rely on the collaboration between public, private, and community-based entities. From expanding digital infrastructure and offering affordable internet options to providing device donation programs and digital literacy training, cities are working toward

Fig. 7.2 Initiatives aimed at bridging the digital divide

creating an inclusive digital environment. These efforts are critical to ensuring that the benefits of AI and smart infrastructure reach all segments of society, particularly those most marginalized by the digital divide.

7.4 Addressing the Digital Divide

The digital divide is one of the most pressing issues in today's technological landscape, exacerbating inequalities in education, healthcare, economic opportunities, and overall societal participation. The divide primarily stems from disparities in access to digital infrastructure, affordable services, and digital literacy. In response, various governments have implemented policy interventions to promote digital literacy and access, aiming to bridge this gap and ensure that all segments of society can participate in the digital economy.

7.4.1 Government Policies on Digital Literacy and Access

The Role of Digital Equity Offices

A pivotal development in addressing the digital divide in the United States has been the establishment of digital equity offices. These offices are created at local, regional, or state levels to ensure that all residents, regardless of income, race, or geographic location, can access digital technologies. The concept of digital equity offices was introduced to address the gaps in broadband access and affordability. For instance, the Brookings Institution highlights that these offices play a critical role in addressing structural barriers that prevent underserved populations from participating in the digital economy (Tomer and Fishbane 2020).

Digital equity offices focus on a variety of measures, including improving broadband infrastructure, negotiating affordable pricing for internet services, and providing digital literacy training programs. These offices collaborate with internet service providers (ISPs), local governments, and community organizations to create inclusive strategies that address the unique needs of disadvantaged communities (Tomer and Fishbane 2020). They also facilitate data collection to monitor progress and ensure accountability, thereby fostering trust between the government and its citizens.

To operationalize the dynamic and inclusive governance role of digital equity offices, a Multi-Agent Reinforcement Learning (MARL) framework is proposed. This model allows multiple digital policy agents (e.g., municipal offices, ISPs, educational entities) to act in a decentralized but coordinated way, learning optimal strategies for resource distribution, pricing negotiations, and community-targeted interventions in real time. MARL enables agents to update their actions continuously based on shared environments and feedback loops from underserved populations, ensuring adaptive digital inclusion policies.

At the core of this framework lies the Multi-Agent Q-learning function, which models the joint learning process of multiple agents in a shared digital policy environment.

$$
\begin{aligned}
Q_i(s, a_1, \ldots, a_n) &\leftarrow Q_i(s, a_1, \ldots, a_n) \\
&\quad + \alpha \left[r_i + \gamma \max_{a'_1, \ldots, a'_n} Q_i\left(s', a'_1, \ldots, a'_n\right) - Q_i(s, a_1, \ldots, a_n) \right]
\end{aligned}
$$

where Q_i is the action-value function for agent I, s is the current state representing socio-economic and digital access indicators, $a_1, \ldots, a_n$ are the joint actions of all agents, r_i is the reward received by agent i (e.g., increase in broadband uptake in a marginalized area), γ is the discount factor for future rewards, s' is the next observed state, and αα is the learning rate that determines how strongly new experiences affect learned values. This function allows each digital equity office, modeled as an autonomous agent, to iteratively improve its decisions such as where to allocate training programs or which partnerships to prioritize by learning from both local outcomes and the coordinated actions of peer agents. The use of the joint action-value function ensures that decisions are not made in isolation but reflect systemic interdependencies and overlapping digital needs across communities.

To enable decentralized yet socially fair policy optimization, we incorporate the Value Decomposition Network (VDN), which aggregates individual agent utilities into a global value function representing collective equity gains.

$$V_{total}\left(s, \overrightarrow{a}\right) = \sum_{i=1}^{n} Q_i(s, a_i)$$

where V_{total} represents the global equity value for state ss and joint action vector $\overrightarrow{a} = (a_1, \ldots, a_n)$, and $Q_i(s, a_i)$ is the value of agent i's action based on its local policy.

This value decomposition function allows the MARL system to balance local optimization with systemic fairness, ensuring that improvements in digital equity are not concentrated but distributed across populations and geographies. It supports the digital equity offices' role in ensuring that no community is disproportionately excluded from digital infrastructure upgrades or literacy initiatives.

To capture evolving citizen feedback and digital inclusion metrics, a Policy Update Function using Proximal Policy Optimization (PPO) is employed, which ensures stable, gradual learning across successive interventions.

$$\mathcal{L}^{CLIP}(\theta) = \mathbb{E}_t[min(r_t(\theta)\hat{A}_t, \text{clip}(r_t(\theta), 1 - \epsilon, 1 + \epsilon)\hat{A}_t)]$$

where $r_t(\theta)$ is the probability ratio between new and old policy, $\hat{A}_t$ is the advantage estimate at time t, ϵ is a small hyperparameter controlling the update range, and $\mathcal{L}^{CLIP}$ is the clipped objective function that prevents large policy shifts, enabling gradual adaptation to citizen needs and data-driven inclusion strategies. This objective function ensures that policy updates remain stable and grounded in real community feedback while enabling the MARL agents to evolve and refine their digital equity interventions over time.

This Multi-Agent Reinforcement Learning model operationalizes the mission of digital equity offices by simulating and optimizing the decisions of multiple urban agents involved in digital inclusion. The model reflects the complex, decentralized nature of digital policy-making across diverse populations and allows for continuous learning based on measurable community outcomes. By integrating Q-learning for adaptive decision-making, value decomposition for fairness, and PPO for stable policy evolution, the model enables cities to implement dynamic, data-driven digital equity strategies that evolve over time making AI a core enabler of fair and inclusive smart urban governance.

Framework: Multi-Agent Reinforcement Learning for Digital Equity Offices (MARL-DEO)
Layered AI Architecture for Decentralized Digital Equity Decision-Making

Layer	Main functionality	AI module(s) and components	Operational scope
Layer 1	Setup: Define socio-economic and digital access indicators for underserved populations	Equity Indicator Profiling Engine	Input → Socio-economic data, community digital access data → Output → Defined indicators for digital equity
Layer 2	Learning: Multi-Agent Q-learning for optimal decisions in digital equity policies	Multi-Agent Q-Learning (MARL)	Input → Agent actions, rewards, digital policy states → Output → Optimized policy decisions for digital equity
Layer 3	Optimization: Value Decomposition Network for balancing local optimization and systemic fairness	Value Decomposition Network (VDN)	Input → Agent utilities → Output → Global value representing collective equity gains
Layer 4	Policy Update: Stable learning using Proximal Policy Optimization (PPO)	Proximal Policy Optimization (PPO)	Input → Policy ratio, advantage estimate → Output → Gradual adaptation of digital equity policies
Layer 5	Evaluation: Monitoring policy effectiveness and real-time adjustments based on outcomes	Adaptive Policy Evaluation Orchestrator	Input → Outcome data, community feedback → Output → Refined policy decisions and interventions

US Digital Strategy and Policy Interventions

The US government has implemented a comprehensive digital strategy through agencies like the US Agency for International Development (USAID) and the Department of State. This strategy includes initiatives such as the Digital Connectivity and Cybersecurity Partnership (DCCP), which focuses on building robust digital infrastructures, promoting secure and interoperable networks, and enhancing digital literacy across regions (Ingram 2021).

The USAID's "Digital Strategy 2020–2025" (USAID 2020) emphasizes the need to promote digital literacy by strengthening digital ecosystems and empowering end-users to effectively utilize digital services. This strategy is particularly important for countries in Southeast Asia and Africa, where access to the internet remains limited. The policy interventions proposed include working with local governments to enact regulatory reforms, offering training programs, and developing digital service platforms that cater to underserved populations (Ingram 2021). These measures aim to reduce the digital divide and promote economic growth in low- and middle-income countries.

To operationalize the equitable implementation of US digital strategy interventions, we propose a Fairness-Constrained Reinforcement Learning (FCRL) model. This model enables an AI agent representing USAID policy mechanisms to learn optimal actions for promoting digital infrastructure, literacy, and platforms across multiple countries, while explicitly ensuring fair distribution of benefits across different socio-demographic groups.

At the core of this framework lies the constrained policy optimization function, which governs how the agent maximizes long-term benefits while satisfying fairness constraints:

$$\pi^* = \underset{\pi E_\pi}{\operatorname{argmax}} \left[\sum\nolimits_{t=0}^{T} \gamma^t R(s_t, a_t) \right] \;\; \text{s.t.} \;\; E_\pi[F_k(s_t, a_t)] \leq \delta_k, \; \forall k$$

where π^* is the optimal policy, π is the policy over time, s_t and a_t are the state and action at time t, $R(s_t, a_t)$ is the reward function representing impact on digital access and literacy, $\gamma \in (0,1)$ is the discount factor, $F_k(s_t, a_t)$ are fairness constraint functions (e.g., disparities in literacy outcomes across genders, regions, or income levels), and δ_k are allowable thresholds for each fairness constraint.

This constrained optimization ensures that digital strategy interventions are not only effective in aggregate but also equitable across the most marginalized groups. The AI agent selects and adapts policy interventions (e.g., launching training platforms or deploying infrastructure) based on evolving local conditions while staying within strict fairness boundaries.

To estimate the advantage of a particular intervention while maintaining compliance with fairness constraints, we introduce the fairness-aware advantage estimation function:

$$\hat{A}_t^{(fair)} = \hat{A}_t - \sum_{k=1}^{K} \lambda_k \nabla_\theta \mathbb{E}_{\pi_\theta}[F_k(s_t, a_t)]$$

where $\hat{A}_t$ is the standard advantage estimate at time t, λ_k are Lagrange multipliers for each fairness constraint k, ∇_θ denotes the gradient with respect to policy parameters θ, and F_k remains the constraint function defined above. This function adjusts the agent's perceived benefit of each action by penalizing decisions that would increase inequity. It ensures that the policy gradient update discourages actions that would, for example, widen digital literacy gaps between urban and rural areas.

To update the policy in a stable and fairness-aware way, we employ the Proximal Policy Optimization (PPO) with Fairness Regularization objective:

$$\mathcal{L}^{PPO+F}(\theta) = \mathbb{E}_t[min(r_t(\theta)\hat{A}_t^{(fair)}, \text{clip}(r_t(\theta), 1 - \epsilon, 1 + \epsilon)\hat{A}_t^{(fair)})]$$

where $r_t(\theta)$ is the probability ratio between new and old policies at time t, ϵ is a small constant controlling update range, and $\hat{A}_t^{(fair)}$ is the fairness-adjusted advantage. This objective ensures the agent updates its policy gradually, avoiding disruptive shifts, while incorporating equity principles directly into learning.

This Fairness-Constrained Reinforcement Learning model allows international digital policy interventions such as those led by USAID to be guided by AI agents that not only learn from local context but also adhere to fairness commitments. By embedding fairness constraints into the reward structure and advantage estimation, the model ensures that decisions made by digital strategy agents are sensitive to the needs of underserved groups without compromising overall impact.

Framework: Fairness-Constrained Reinforcement Learning for Digital Strategy Interventions (FCRL-DSI)

Layered AI Architecture for Equitable Digital Policy Interventions

Layer	Main functionality	AI module(s) and components	Operational scope
Layer 1	Setup: Define socio-economic and digital access indicators for underserved regions	Equity Indicator Definition Engine	Input → Socio-economic data, digital access indicators → Output → Defined criteria for equity-focused interventions
Layer 2	Learning: Fairness-constrained decision-making for digital policy interventions using reinforcement learning	Fairness-Constrained Reinforcement Learning (FCRL)	Input → Policy decisions, socio-economic data, community feedback → Output → Optimized interventions with fairness constraints
Layer 3	Optimization: Fairness-aware advantage estimation to guide policy updates with equity consideration	Fairness-Aware Advantage Estimation Module	Input → Policy feedback, fairness constraints → Output → Fairness-adjusted advantage values for policy updates
Layer 4	Policy Update: Stable policy learning with fairness regularization using Proximal Policy Optimization (PPO)	Proximal Policy Optimization (PPO)	Input → Policy ratios, fairness-adjusted advantage → Output → Gradual policy evolution incorporating fairness
Layer 5	Evaluation: Monitoring and evaluation of policy outcomes, ensuring fairness and impact measurement	Policy Impact Evaluation Engine	Input → Outcome data, community feedback → Output → Refined policy decisions and ongoing intervention adjustments

Libraries as Catalysts for Digital Literacy

Libraries have been recognized as critical actors in fostering digital literacy and inclusion. Wang and Si (2024) argue that libraries play a central role in promoting equitable access to information and digital skills. In China, for example, public libraries have implemented a variety of programs to train individuals in digital literacy, emphasizing the need for public access to technology, particularly in rural and underserved areas.

Public policy has supported these initiatives by encouraging libraries to offer training programs that focus on digital competencies, internet navigation, and the use of digital tools for education, employment, and civic engagement (Wang and Si 2024). These programs are often backed by government-funded initiatives that provide libraries with the necessary infrastructure and resources to serve as community hubs for digital education.

To empower libraries as decentralized hubs for digital education, we propose a Federated Learning (FL) framework. FL allows each library to train its own local AI model based on community interactions and literacy program engagement, without sharing raw user data. These local models are periodically aggregated to improve a shared global model that reflects collective learning while respecting local privacy and heterogeneity.

At the core of this framework lies the Federated Averaging (FedAvg) function, which defines how local model updates from libraries are aggregated into a global literacy model:

$$w_t^{(global)} = \sum_{k=1}^{K} \frac{n_k}{n} w_t^{(k)}$$

where $w_t^{(global)}$ is the aggregated global model at training round t, $w_t^{(k)}$ is the locally trained model at library k, n_k is the number of training examples at library k, and $n = \sum_{k=1}^{K} n_k$ is the total number of training examples across all libraries. This function allows each library to contribute proportionally to the shared model based on the volume of its training data, ensuring that large and small libraries influence the global literacy AI model in a balanced way.

To adapt each local library model to its specific context, we define the local training objective function for each library node:

$$\min_{w} f_k(w) = \frac{1}{n_k} \sum_{i=1}^{n_k} \ell(w; x_i^{(k)}, y_i^{(k)})$$

where $f_k(w)$ is the local loss function at library k, $\ell(w; x_i^{(k)}, y_i^{(k)})$ is the loss for a single training example $(x_i^{(k)}, y_i^{(k)})$, and ww represents the model parameters. This function allows each library to train its AI literacy model on local digital engagement data such as usage frequency of e-learning tools, types of queries made, or user completion of digital modules without exposing raw data to the central server.

To ensure that the global model improves over time while minimizing communication overhead and respecting bandwidth limits (particularly in rural libraries), we apply round-based asynchronous updates governed by the global convergence condition:

$$\Delta_t = \|w_t^{(global)} - w_{t-1}^{(global)}\|_2 < \epsilon$$

where Δ_t is the change in the global model between training rounds t and $t - 1$, and ϵ is a small threshold ensuring convergence. This function ensures that global model updates proceed only when significant learning improvements are observed, reducing redundant data exchange between central servers and libraries.

This Federated Learning model transforms libraries into intelligent, collaborative nodes for digital literacy promotion. Each library continuously adapts its AI training modules based on local data such as the digital skills needs of rural versus urban populations while contributing to a shared, privacy-preserving global model of educational advancement. The Federated Averaging function ensures equitable participation, the local training objective captures region-specific learning dynamics, and the convergence condition ensures communication efficiency even in bandwidth-constrained regions. This AI-driven architecture reflects the core educational and infrastructural mission of public libraries, aligning with public policy mandates while enhancing their impact through adaptive, decentralized intelligence.

Framework: Federated Learning for Digital Literacy Promotion in Libraries (FL-DLP)
Layered AI Architecture for Decentralized Digital Literacy Training

Layer	Main functionality	AI module(s) and components	Operational scope
Layer 1	Setup: Define socio-economic and digital engagement data for underserved communities	Digital Literacy Indicator Definition Engine	Input → Socio-economic data, community interaction data → Output → Defined indicators for digital literacy interventions
Layer 2	Learning: Federated Learning for decentralized training of literacy models across libraries	Federated Learning (FL), Federated Averaging (FedAvg)	Input → Local data from libraries → Output → Aggregated global model for digital literacy
Layer 3	Optimization: Local model training using specific library engagement data	Local Model Training Module	Input → Local engagement data from libraries → Output → Locally trained literacy models

(continued)

(continued)

Layer	Main functionality	AI module(s) and components	Operational scope
Layer 4	Monitoring: Monitoring model convergence and ensuring communication efficiency for updates	Federated Communication Monitoring Engine	Input → Global model updates, communication feedback → Output → Refined global model with minimal communication overhead
Layer 5	Evaluation: Evaluate global model performance and adapt training strategies for local conditions	Adaptive Model Evaluation Module	Input → Model evaluation metrics, library feedback → Output → Adjusted training strategies and refined models

Addressing Regional Disparities in Digital Access

Regional disparities in access to digital technology remain a significant issue, particularly in large and diverse countries like China. Policies aimed at addressing these disparities often focus on improving digital infrastructure in rural areas and providing affordable internet access to low-income populations (Wang and Si 2024). Government initiatives such as the "Action Plan for Enhancing National Digital Literacy and Skills" have been instrumental in reducing these regional inequalities by promoting internet access and digital education across different regions.

In addition to infrastructure improvements, these policies also focus on enhancing digital skills through targeted educational programs, often delivered through community organizations or public institutions like libraries. Such programs are critical in ensuring that individuals not only have access to the internet but also the skills to navigate and utilize digital platforms effectively (Wang and Si 2024).

To support national efforts in addressing regional digital disparities, we propose a Geographically Weighted Multi-task Learning (GW-MTL) model. This model allows policymakers to learn multiple correlated digital access outcomes (such as internet availability, affordability, and literacy skills) while weighting these tasks based on region-specific socioeconomic, infrastructural, and geographic characteristics.

At the core of this framework lies the multi-task objective function with spatial weighting, which enables the model to learn predictive tasks across regions while respecting spatial heterogeneity:

$$\min_{\Theta}\sum\nolimits_{r=1}^{R} w_r \sum\nolimits_{t=1}^{T} \mathcal{L}_t^{(r)}\left(f_t^{(r)}\left(X^{(r)};\theta_t\right), Y_t^{(r)}\right) + \Omega(\Theta)$$

where R is the number of regions, T is the number of prediction tasks, w_r is the spatial weight assigned to region r, $\mathcal{L}_t^{(r)}$ is the loss function for task t in region r, $f_t^{(r)}$ is the region-specific prediction function for task t, $X^{(r)}$ and $Y_t^{(r)}$ are the input features and target outputs for region r, θ_t are the task-specific model parameters, and $\Omega(\Theta)$ is a regularization term controlling smoothness across tasks. This formulation enables the model to learn multiple outcomes concurrently such as broadband coverage, digital affordability, and user skills while modulating the contribution of each region through spatial weights that reflect geographic, infrastructural, and economic proximities.

To compute these spatial weights, the model employs a Gaussian kernel based on the distance between the focal region and its neighbors, thereby ensuring that geographically or socioeconomically similar regions exert greater influence in the learning process:

$$w_r = \exp\left(-\frac{d(r, r_0)^2}{2\sigma^2}\right)$$

where $d(r, r_0)^2$ is the Euclidean distance between region r and a reference region r_0, and σ is the kernel bandwidth controlling the rate at which regional influence decays with distance.

This function ensures that prediction for a given region is not driven solely by national trends but is contextually informed by localized realities, particularly useful in vast and diverse territories such as China where digital infrastructure and literacy vary widely across provinces.

To capture the latent relationships between prediction tasks such as the interplay between internet availability and literacy outcomes the model introduces a regularization function that enforces parameter similarity across related tasks:

$$\Omega(\Theta) = \lambda \sum_{t<t'} \|\theta_t - \theta_{t'}\|_2^2$$

where λ is the regularization strength, θ_t and $\theta_{t'}$ are the parameter vectors of tasks t and t', and $\|\theta_t - \theta_{t'}\|_2^2$ encourages alignment across task-specific models.

This term enhances the model's ability to generalize by ensuring that tasks which are empirically or conceptually related such as access and affordability benefit from shared representational structures while still allowing task-specific adaptations where regional deviations are substantial.

The Geographically Weighted Multi-task Learning model enables governments and planners to model, predict, and address regional digital disparities with scientific precision. It allows simultaneous forecasting of multiple interdependent digital outcomes while adjusting for spatial proximity and socioeconomic similarity. This approach provides an AI-driven basis for prioritizing policy interventions such as where to invest in infrastructure or deploy literacy programs with maximal regional impact. By combining spatial weighting, task coupling, and local prediction fidelity, GW-MTL transforms how digital equity policies are designed, evaluated, and refined across large heterogeneous territories.

Framework: Geographically Weighted Multi-Task Learning for Regional Digital Equity (GW-MTL-RDE)

Layered AI Architecture for Decentralized Digital Equity Policy Interventions

Layer	Main functionality	AI module(s) and components	Operational scope
Layer 1	Setup: Define regional socio-economic, infrastructural, and digital access indicators	Regional Indicator Definition Engine	Input → Socio-economic data, regional digital access data → Output → Defined regional indicators for digital equity interventions
Layer 2	Learning: Geographically Weighted Multi-task Learning (GW-MTL) for region-specific predictions	Geographically Weighted Multi-task Learning (GW-MTL)	Input → Regional data, multiple tasks (e.g., broadband, affordability, literacy) → Output → Region-specific predictions and recommendations
Layer 3	Optimization: Calculate spatial weights using a Gaussian kernel to account for regional proximity	Spatial Weight Computation Module	Input → Geographic data, region distance → Output → Spatial weight values for each region
Layer 4	Regularization: Enforce parameter similarity across related tasks using a regularization function	Cross-Task Regularization Engine	Input → Task-specific model parameters → Output → Regularized task parameters ensuring shared knowledge
Layer 5	Evaluation: Evaluate model performance and adapt strategies for local digital equity interventions	Regional Strategy Adaptation Module	Input → Model performance metrics, regional feedback → Output → Adapted digital equity strategies and interventions

Policy Impact and Future Directions

The implementation of digital literacy policies has already shown promising results in reducing the digital divide in many regions. For instance, the establishment of digital equity offices in US cities like Boston and San Francisco has led to improved broadband access for low-income households, and libraries have played an instrumental role in fostering digital literacy in underserved communities (Tomer and Fishbane 2020).

However, challenges remain, particularly in ensuring the sustainability of these initiatives. Funding for digital equity programs is often limited, and the rapid pace of technological advancement means that policies must continually evolve to address new challenges such as cybersecurity risks, data privacy concerns, and the need for advanced digital skills (Ingram 2021).

To support the long-term sustainability and adaptability of digital equity interventions, we propose a Bayesian Reinforcement Learning (BRL) framework. This model enables policy agents to learn optimal funding and program adaptation strategies under conditions of uncertainty, integrating both historical policy outcomes and probabilistic reasoning about future digital challenges. BRL is particularly suited for this domain because it explicitly models both the expected utility of policy decisions and the uncertainty associated with those expectations allowing policymakers to make informed trade-offs between exploiting known interventions and exploring new ones.

At the heart of this model lies the Bayesian value function with posterior belief updating, which determines the expected long-term return of a given action aa in state ss while accounting for uncertainty in the underlying transition dynamics:

$$V^{\pi}(s) = \mathbb{E}_{P \sim \mathcal{B}}[\sum_{t=0}^{T} \gamma^t R(s_t, a_t) \mid \pi, P]$$

where $V^{\pi}(s)$ is the expected value of following policy π from state s, $R(s_t, a_t)$ is the reward for taking action a_t in state s_t, $\gamma \in (0,1)$ is the discount factor, $P \sim \mathcal{B}$ denotes that the environment transition probabilities P are sampled from a Bayesian posterior distribution $\mathcal{B}$, and the expectation is over all possible environment models consistent with current beliefs. This formulation allows decision-makers to continuously update their understanding of how funding and programmatic choices (e.g., investing in cybersecurity training or upgrading access infrastructure) impact outcomes, while explicitly quantifying the uncertainty of these effects. It also supports learning in environments where historical data may be sparse or only partially observable, as is common in early-stage digital equity deployments.

To update the agent's belief about the environment dynamics as new policy outcomes are observed, we define the Bayesian posterior update for the transition model:

$$\mathcal{B}_{t+1}(P) \propto \mathcal{B}_t(P) \cdot P(s_{t+1} \mid s_t, a_t)$$

where $\mathcal{B}_t(P)$ is the agent's belief over transition probabilities at time t, $s_{t+1} \mid s_t, a_t$ is the observed state transition after taking action atat, and $\mathcal{B}_{t+1}(P)$ is the updated belief after incorporating the new observation. This update mechanism allows the model to refine its understanding of how specific interventions perform over time, accounting for shifting digital needs, evolving technologies, and changing user behavior.

To decide which action to take at each step—balancing the desire to optimize known high-impact strategies with the need to explore potentially better alternatives—we introduce a Bayesian action-selection function based on Thompson Sampling:

$$a_t = \underset{a \in A}{\arg max} \mathbb{E}_{P \sim \mathcal{B}_t}[Q(s_t, a; P)]$$

where a_t is the action selected at time t, A is the action space (e.g., fund allocation options or strategic pivots), $Q(s_t, a; P)$ is the action-value function under transition model P, and $\mathbb{E}_{P \sim \mathcal{B}_t}$ represents the expectation over sampled transition dynamics from the current belief. This function ensures that policy decisions are not just greedy with respect to expected value but also exploratory with respect to model uncertainty, which is essential when confronting future unknowns like emerging digital threats.

This Bayesian Reinforcement Learning model provides a scientifically grounded framework for evaluating and adapting digital literacy policies over time. By integrating posterior uncertainty into both the value estimation and decision process, the model allows policy agents to reason probabilistically about future risks, respond intelligently to new data, and optimize impact even when resources are constrained. It directly supports sustainability by enabling dynamic policy refinement, and it equips institutions to respond to technological change, cyber risk, and evolving skill demands through robust, adaptive learning. In doing so, BRL transforms digital governance from static policy execution into an intelligent, evidence-based cycle of observation, inference, and strategic adaptation.

Framework: Bayesian Reinforcement Learning Framework for Digital Literacy Policies (BRL-DLP)

Layered AI Architecture for Sustainable Digital Equity Interventions

Layer	Main functionality	AI module(s) and components	Operational scope
Layer 1	Data Collection and Analysis: Collect and analyze data from previous digital literacy policies	Policy Data Preprocessing Engine	Input → Historical data, policy outcomes → Output → Processed data for AI model
Layer 2	Bayesian Value Estimation and Belief Update: Update belief about transition probabilities and expected outcomes	Bayesian Reinforcement Learning (BRL)	Input → Historical policy outcomes, transition model → Output → Updated belief state
Layer 3	Action Selection: Select optimal action balancing known and exploratory strategies	Thompson Sampling	Input → Updated belief state, action space → Output → Selected action
Layer 4	Policy Refinement and Adjustment: Refine policies based on feedback and real-time data to ensure long-term sustainability	Adaptive Policy Refinement Module	Input → Monitored outcomes, evolving data → Output → Adjusted policy and refined strategy

Figure 7.3 illustrates government policies on digital literacy and access through five interconnected areas: the role of digital equity offices, US digital strategy, libraries as digital literacy hubs, reducing regional disparities, and policy impact and future challenges. Digital equity offices aim to ensure universal digital access across demographics by improving broadband availability, affordability, and digital literacy, while collaborating with ISPs and community organizations. The US digital strategy focuses on building infrastructure and secure networks, such as DCCP, promoting digital literacy in underserved regions through initiatives such as USAID, and supporting national digital inclusion goals. Libraries are presented as digital literacy hubs that provide access to technology and digital skills training, especially in rural and underserved areas, with government support. Reducing regional disparities involves expanding infrastructure and affordable access in rural areas, promoting digital skills through local programs, and targeting regional gaps in digital inclusion. Policy impact and future challenges include improved access and literacy, while funding, adaptability, and long-term implementation challenges remain. The figure presents government policy as a coordinated framework for expanding digital access, strengthening digital literacy, reducing regional inequalities, and sustaining inclusive digital participation.

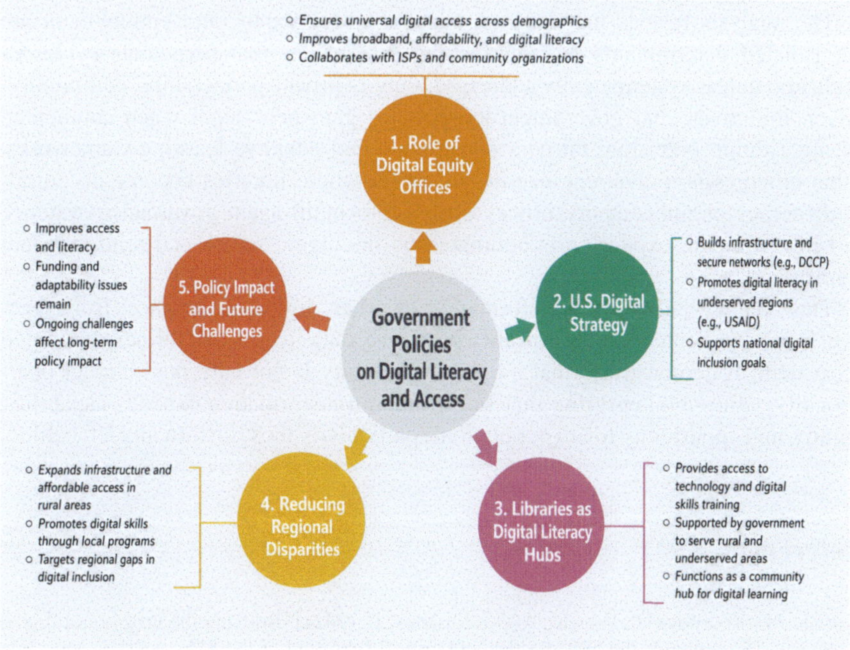

Fig. 7.3 Government policies on digital literacy and access

Government policies aimed at promoting digital literacy and access are crucial for addressing the digital divide. Initiatives such as the creation of digital equity offices, the implementation of national digital strategies, and the support of libraries as digital literacy hubs are all critical components of a comprehensive approach to bridging the digital divide. These policies not only provide the necessary infrastructure and access but also ensure that individuals have the skills to participate fully in the digital economy.

7.5 Conclusion

This chapter demonstrates that the digital divide is neither an abstract concept nor a passive consequence of technological evolution; it is a structural inequity that must be actively dismantled through measurable, optimizable, and governable strategies. By introducing a Digital Equity Index and embedding it within advanced AI architectures ranging from reinforcement learning for subsidy allocation and digital literacy, to Graph Neural Networks for mapping infrastructural disparities, to fairness-constrained policy optimization, it reframes digital inclusion as an enforceable guarantee rather than a charitable aspiration.

The analysis reveals that income, education, geography, and employment are not isolated determinants but interlocking dimensions that perpetuate cycles of exclusion unless systematically addressed. Public–private partnerships, community-driven initiatives, and government-led policies gain new depth when augmented by algorithmic precision, fairness constraints, and adaptive learning frameworks. What emerges is a civic engineering blueprint where libraries become decentralized literacy engines, equity offices evolve into multi-agent governance systems, and infrastructure expansion is optimized by intelligent search and reinforcement algorithms.

This chapter calls for a transformation of urban digital governance: from reactive, fragmented interventions toward a proactive, data-driven, and fairness-anchored ecosystem. It demonstrates that a digitally just city is not only possible but operationally achievable, ensuring that smart urban transformation expands capability, dignity, and opportunity for all populations, particularly those historically left behind.

References

Feurich, M., Kourilova, J., Pelucha, M., & Kasabov, E. (2024). Bridging the urban-rural digital divide: Taxonomy of the best practice and critical reflection of the EU countries' approach. European Planning Studies. https://doi.org/10.1080/09654313.2023.2186167

Ingram, G. (2021). Bridging the global digital divide: A platform to advance digital development in low- and middle-income countries. *Brookings Institution.* https://www.brookings.edu/wp-content/uploads/2021/05/Bridging-the-Digital-Divide_final.pdf

San Francisco Digital Equity Strategic Plan 2019–2024. https://media.api.sf.gov/documents/SF_Digital_Equity_Strategic_Plan_2019_EG9ePTv.pdf

Tomer, A., & Fishbane, L. (2020). Bridging the digital divide through digital equity offices. *Brookings Metropolitan Policy Program.* https://www.brookings.edu/articles/bridging-the-digital-divide-through-digital-equity-offices/

USAID, 2020. *USAID Digital Strategy 2020–2024*, USAID: United States Agency for International Development. United States of America. Retrieved from https://coilink.org/20.500.12592/7wm6n60 on 03 Oct 2025. COI: 20.500.12592/7wm6n60.

Wang, C., & Si, L. (2024). The intersection of public policy and public access: Digital inclusion, digital literacy education, and libraries. *Sustainability, 16.* https://doi.org/10.3390/su16051878

Chapter 8
AI and Equity by Design: Bias Mitigation, Fairness, and Human-Centered Frameworks

Artificial intelligence is no longer a neutral instrument; it is a force that can either cement historical inequities or actively remediate them. This chapter, AI and Equity by Design, contends that fairness cannot be retrofitted at the margins of technical pipelines; it must be architected into the very objectives, losses, constraints, and monitoring loops that govern modern AI. We map, with precision, how bias arises from skewed datasets and design shortcuts to the subtle leakage of human prejudice into learned representations, and we demonstrate how these failure modes manifest in high-stakes arenas such as healthcare, criminal justice, education, and employment. The result is a rigorous, system-level diagnosis paired with actionable remedies that elevate equity from aspiration to operational requirement.

The chapter proceeds from measurement to mitigation to accountability. First, it formalizes bias detection using established fairness metrics and stress-tests models across protected groups to expose disparities that typical accuracy scores conceal. Second, it introduces fairness-aware learning strategies: adversarial debiasing, constraint-based optimization, and synthetic rebalancing via conditional generative models that suppress discriminatory signal paths without sacrificing predictive power. Third, it advances human-centered AI as an optimization program: multi-stakeholder utilities are aggregated and explicitly regularized so that system performance is inseparable from social benefit, with continuous post-deployment auditing to prevent fairness decay in the wild.

Crucially, the chapter binds technical integrity to ethical accountability. It integrates explainability (e.g., attribution methods that decompose decisions feature-by-feature), developer-traceable optimization trails, counterfactual stability checks, and governance-ready reporting so that every consequential output can be interrogated, attributed, and corrected. Readers will find layered reference architectures that translate principles into reproducible practice from bias discovery and debiasing to monitoring, rollback, and reinforcement of responsible behavior. By uniting mathematical rigor with a humanistic imperative, the chapter offers an international roadmap for

D. Sargiotis, *Empathy in Algorithms*, https://doi.org/10.1007/978-3-032-32128-2_8

building AI systems that are not only accurate and scalable, but transparent, just, and worthy of public trust.

8.1 Sources of Bias in AI Systems

Bias in AI systems often originates from multiple sources, including biased datasets, algorithmic design flaws, and the unintentional introduction of human biases during model training. According to Ferrara (2023), AI models can learn and replicate patterns of bias present in the data used to train them. This has led to discriminatory outcomes in areas such as facial recognition, hiring algorithms, and criminal justice systems. For instance, facial recognition systems have been shown to be less accurate for individuals with darker skin tones, leading to higher rates of false positives (Ferrara 2023). To identify and mitigate the impact of bias in AI systems, we introduce a dual-layered fairness-aware modeling framework that combines statistical bias detection with real-time adversarial correction. This model first detects disparity across protected groups using a Statistical Parity Difference (SPD) function and then applies Adversarial Debiasing to retrain the model in a fairness-constrained adversarial setting. The Statistical Parity Difference (SPD) is defined as:

$$\mathrm{SPD} = P(\hat{y} = 1 \mid A = 0) - P(\hat{y} = 1 \mid A = 1)$$

where A is the protected attribute (e.g., skin tone, gender, disability status), and $\hat{y}$ is the AI system's predicted outcome (e.g., match score in facial recognition, interview recommendation). SPD measures the difference in positive classification rates between unprivileged ($A = 1$) and privileged ($A = 0$) groups. A non-zero SPD indicates systemic bias: the larger the deviation, the more discriminatory the system is.

To address this disparity, we incorporate an Adversarial Debiasing Framework, where a primary predictor model is trained in tandem with an adversary that attempts to predict the protected attribute AA from the model's predictions. The predictor seeks to maximize performance, while the adversary tries to uncover bias. The final objective function balances these two goals:

$$\underset{\theta_f}{min}[\mathcal{L}_{\text{task}}(\hat{y}, y) - \lambda \cdot \underset{\theta_a}{max}\mathcal{L}_{\text{adv}}(A, \hat{y})]$$

where $\mathcal{L}_{\text{task}}(\hat{y}, y)$ is the task loss between predicted and actual outcomes, $\mathcal{L}_{\text{adv}}(A, \hat{y})$ is the adversarial loss quantifying how well group membership A can be inferred from predictions $\hat{y}$, θ_f and θ_a are the parameters of the predictor and adversary respectively, and λ controls the strength of the fairness constraint. The adversary is trained to maximize bias detection, while the predictor learns to produce fair predictions that conceal protected group identifiers.

Together, these two components form a closed-loop feedback system in which bias is not only identified using statistical metrics like SPD but is dynamically mitigated during model training.

This integrated model illustrates how bias in AI systems can be rigorously measured, evaluated, and corrected using formal mathematical strategies. In domains such as facial recognition, hiring, or criminal justice, unmitigated bias can amplify structural inequalities. By embedding fairness directly into the optimization process through adversarial learning, this model ensures that AI systems do not merely avoid discriminatory outcomes after deployment, but learn to avoid them during their formation. Such frameworks are essential for building public trust, regulatory compliance, and ethical robustness in AI-driven public infrastructures.

Framework: Fairness-Aware Adversarial Learning Framework for Bias Mitigation (FAAL-BM)

Layered AI Architecture for Mitigating Bias in AI Systems

Layer	Main functionality	AI module(s) and components	Operational scope
Layer 1	Bias Detection: Detect disparity across protected groups using Statistical Parity Difference (SPD)	Bias Disparity Detection Engine	Input → Predictions from model, protected group labels → Output → SPD value indicating bias
Layer 2	Adversarial Debiasing: Retrain model in an adversarial setting to reduce bias	Adversarial Debiasing	Input → Model predictions, protected group labels → Output → Fairer predictions with reduced bias
Layer 3	Bias Correction: Apply fairness constraints during model training to ensure fairness	Fairness Constraint Integration Module	Input → Adversarially trained model, fairness constraints → Output → Model with balanced fairness
Layer 4	Continuous Monitoring: Monitor fairness post-deployment and adjust model when necessary	Adaptive Fairness Monitoring Engine	Input → Post-deployment outcomes, feedback → Output → Adjusted model with continuous fairness monitoring

In healthcare, AI systems have been found to reinforce existing disparities. For example, certain predictive algorithms trained on biased datasets have incorrectly assessed the health risks of minority groups, as seen in studies where Black patients were disproportionately assigned lower risk scores despite having higher medical needs (Ferrara 2023). Such biases in AI can exacerbate systemic inequalities and lead to unfair treatment.

To mitigate discriminatory outcomes in AI-driven healthcare systems, we propose a Fairness-Constrained Adversarial Debiasing (FCAD) framework. This model jointly optimizes predictive accuracy and fairness by training two competing networks: a predictor that estimates health risk scores and an adversary that attempts to infer sensitive attributes (such as race) from the predictor's output. By minimizing the predictor's loss while maximizing the adversary's loss, the model ensures that risk predictions are informative yet statistically independent from protected characteristics. Additionally, a fairness constraint such as Equalized Odds is applied to ensure outcome parity across demographic groups.

The core training objective of the FCAD model is expressed as a minimax optimization problem:

$$\min_{\theta_P} \max_{\theta_A} \mathcal{L}_{pred}\left(Y, \hat{Y}_{\theta_P}\right) - \lambda \cdot \mathcal{L}_{adv}\left(A, \hat{A}_{\theta_P, \theta_A}\right)$$

where $\mathcal{L}_{pred}(Y, \hat{Y}_{\theta_P})$ is the loss function for the predictor estimating the true outcome Y (e.g., health risk score), Y, $\hat{Y}_{\theta_P}$ is the predictor's output, $\mathcal{L}_{adv}(A, \hat{A}_{\theta_P, \theta_A})$ is the adversarial loss that tries to predict the protected attribute A (e.g., race) from the predictor's output using adversarial parameters θ_A, λ is a non-negative hyperparameter balancing fairness and accuracy, and θ_P are the predictor's parameters. This formulation drives the predictor to produce outputs that are both accurate and uninformative of sensitive features, thereby discouraging discriminatory behavior embedded in biased training data. The adversary simulates a fairness auditor: if it can easily guess race from the prediction, the model is penalized. This interaction improves the predictor's robustness against encoding bias in its latent representations.

To enforce a formal fairness constraint, we apply the Equalized Odds condition during training by minimizing the difference in true positive and false positive rates across groups. This is quantified by a fairness penalty:

$$\mathcal{L}_{eo} = \left|\Pr\left(\hat{Y} = 1|Y = 1, A = 0\right) - \Pr\left(\hat{Y} = 1|Y = 1, A = 1\right)\right| +$$

$$|Pr(\hat{Y} = 1 \mid Y = 0, A = 0) - Pr(\hat{Y} = 1 \mid Y = 0, A = 1)|$$

where $Pr(\hat{Y} = 1 \mid Y = 1, A = a)$ and $Pr(\hat{Y} = 1 \mid Y = 0, A = a)$ are the true and false positive rates for demographic group $A = a$, with $A = 0$ and $A = 1$ indicating different groups (e.g., racial categories). The goal is to equalize these rates, thereby removing systemic advantages or disadvantages for any particular group in prediction outcomes.

The total loss function for FCAD thus becomes a combined objective that integrates predictive accuracy, adversarial fairness enforcement, and explicit fairness constraint regularization:

$$\mathcal{L}_{total} = \mathcal{L}_{pred} - \lambda \cdot \mathcal{L}_{adv} + \mu \cdot \mathcal{L}_{eo}$$

where μ is the penalty weight for the Equalized Odds constraint, ensuring that the final model satisfies both adversarial independence and fairness in statistical outcomes.

This Fairness-Constrained Adversarial Debiasing model provides a rigorous, real-time correction mechanism for biased risk prediction in healthcare AI systems. It enables AI to maintain predictive power while actively suppressing discriminatory signal paths linked to race, gender, or other protected characteristics. By enforcing Equalized Odds, the model guarantees that medical decisions are made equitably, regardless of demographic identity. This architecture addresses the exact type of systemic failure highlighted in the original section: disproportionate underestimation of health risks for minority patients. FCAD ensures that AI models deployed in healthcare not only perform well but also uphold ethical and social standards of justice in algorithmic treatment pathways.

Framework: Fairness-Constrained Adversarial Debiasing Framework for Healthcare Bias Mitigation (FCAD-HBM)

Layered AI Architecture for Mitigating Discriminatory Outcomes in Healthcare AI Systems

Layer	Main functionality	AI module(s) and components	Operational scope
Layer 1	Bias Detection: Detect disparity in model predictions using fairness metrics (e.g., Equalized Odds)	Fairness Metric Evaluation Engine	Input → Predictions from model, protected group labels → Output → Fairness metrics indicating bias
Layer 2	Adversarial Debiasing: Train model with adversarial network to reduce bias and improve fairness	Adversarial Debiasing	Input → Model predictions, protected group labels → Output → Fairer predictions with reduced bias
Layer 3	Fairness Constraint Enforcement: Apply Equalized Odds to ensure parity in outcome rates across groups	Equalized Odds Enforcement Module	Input → Model predictions, fairness constraints → Output → Outcome parity between groups
Layer 4	Continuous Monitoring: Monitor fairness post-deployment and adjust model based on real-time feedback	Real-Time Fairness Monitoring Engine	Input → Post-deployment outcomes, feedback → Output → Adjusted model ensuring continuous fairness

8.2 The Role of Human-Centered AI

Human-centered AI (HCAI) provides a framework for addressing bias by involving diverse stakeholders in the development and deployment of AI systems. Chen et al. (2023) advocate for a human-centered approach, emphasizing the need to consider social, ethical, and cultural contexts throughout the AI lifecycle. HCAI integrates the perspectives of end-users, ethicists, social scientists, and engineers to ensure that AI systems are not only effective but also fair and equitable (Chen et al. 2023). By prioritizing human well-being and focusing on the specific needs of marginalized communities, HCAI can prevent AI from reinforcing existing societal inequalities.

To systematically operationalize the principles of human-centered AI (HCAI), we introduce a Multi-Stakeholder System Alignment Model that embeds ethical diversity into the AI learning process. This framework ensures that the optimization of AI behavior does not prioritize narrow technical efficiency over broad human values. The model is structured in two layers: the first formalizes diverse stakeholder priorities into a unified optimization goal; the second imposes fairness constraints to prevent disproportionate harm to vulnerable populations. The Stakeholder Utility Aggregation Function is defined as:

$$U_{\text{HCAI}}(a_t) = \sum_{k=1}^{K} \omega_k \cdot u_k(a_t)$$

where $u_k(a_t)$ is the perceived utility of action a_t for stakeholder group k, including end-users, ethicists, engineers, marginalized communities, and social scientists, ω_k is the normalized importance weight of each group (with $\sum \omega_k = 1$), and K is the number of stakeholder groups involved. This function aggregates diverse preferences into a unified expression of human-centered system value.

The final human-centered AI optimization objective becomes:

$$\max_{a_t}[U_{\text{HCAI}}(a_t) - \lambda \cdot \mathcal{L}_{\text{ineq}}(a_t)]$$

where λ is a tunable fairness sensitivity coefficient that controls the trade-off between maximizing aggregate utility and minimizing inequality. This formulation allows human-centered AI (HCAI) systems to learn from multi-stakeholder input, generate socially aligned outputs, and avoid outcomes that benefit some users at the expense of others.

To ensure that system optimization does not disproportionately benefit dominant groups, we integrate a Fairness Penalty Function grounded in equality of benefit:

$$\mathcal{L}_{\text{ineq}}(a_t) = \max_{k,l}|u_k(a_t) - u_l(a_t)|$$

This loss penalizes inequality in stakeholder satisfaction across all group pairs k, l. The system thereby learns to minimize the gap in perceived value delivered to each group, particularly prioritizing equity for those historically excluded from design processes.

This model brings mathematical structure to the core of human-centered AI as advocated by Chen et al. (2023). It formalizes the ethical principle that fairness is not a post-processing concern but a design-time constraint. Through dynamic aggregation of diverse stakeholder perspectives and explicit equity penalties, AI systems become capable of adapting their behavior to serve not just the average user but the most vulnerable. This architecture provides the foundation for building intelligent systems that are not only accurate and scalable but also socially reflexive, inclusive, and just.

Framework: Human-Centered AI Multi-Stakeholder System Alignment Model (HCAI-MSSAM)

Layered AI Architecture for Ethical AI and Fairness in Multi-Stakeholder Systems

Layer	Main functionality	AI module(s) and components	Operational scope
Layer 1	Data Collection and Stakeholder Preferences: Gather diverse stakeholder input and formalize priorities	Stakeholder Preference Aggregation Engine	Input → Stakeholder preferences (e.g., end-users, ethicists, marginalized communities) → Output → Aggregated stakeholder priorities
Layer 2	Utility Aggregation: Aggregate stakeholder utility preferences into a unified optimization goal	Utility-Based Multi-Stakeholder Decision Making	Input → Stakeholder preferences → Output → Unified utility function representing human-centered values
Layer 3	Fairness Constraint Enforcement: Apply fairness constraints to ensure equity in stakeholder satisfaction	Fairness-Aware Loss Functions	Input → Predicted outputs, fairness constraints → Output → Fairness penalty to equalize stakeholder satisfaction
Layer 4	Continuous Monitoring: Monitor system outcomes and adjust to ensure long-term fairness	Fairness Alignment Monitoring Module	Input → Post-deployment feedback, real-time data → Output → Adjusted model ensuring continuous fairness alignment

8.3 Mitigating Bias in AI

Several strategies have been proposed to mitigate bias in AI, including enhancing data diversity, improving algorithm transparency, and using bias-auditing tools. According to Al-kfairy et al. (2024), AI models must be trained on diverse datasets that represent the full spectrum of the population they serve. This involves collecting data from underrepresented groups and using synthetic data generation techniques to ensure equitable representation (Al-kfairy et al. 2024). To operationalize effective bias mitigation in AI systems, we propose a Fairness-Aware Training Pipeline that integrates dataset diversification with constraint-based optimization and post-hoc fairness auditing. The pipeline begins with the use of synthetic data generation via Generative Adversarial Networks (GANs) to rebalance training datasets where underrepresented demographic groups lack sufficient representation. This is followed by training a predictive model under a bias-aware loss function, and final validation is performed using interpretable fairness metrics such as the Disparate Impact Ratio (DIR). To address underrepresentation during data preprocessing, synthetic rebalancing is performed via Conditional GANs (cGANs):

$$x_{\text{synthetic}} \sim G(z \mid A = a)$$

where $G(z \mid A = a)$ generates synthetic data conditioned on a protected attribute $A = a$, with z as the latent input noise vector. This ensures that underrepresented classes are realistically populated without compromising data integrity or privacy.

Next, we define the Bias-Regularized Loss Function as:

$$\mathcal{L}_{\text{total}} = \mathcal{L}_{\text{task}} + \lambda \cdot \mathcal{L}_{\text{bias}}$$

where $\mathcal{L}_{\text{task}}$ is the original prediction loss (e.g., cross-entropy for classification), $\mathcal{L}_{\text{bias}}$ quantifies the disparity in predictions across protected attributes (e.g., ethnicity, gender, disability), and λ is a regularization hyperparameter controlling the sensitivity to fairness violations.

To ensure demographic parity, $\mathcal{L}_{\text{bias}}$ is defined using the squared difference in positive prediction rates:

$$\mathcal{L}_{\text{bias}} = \sum_{g \in G} \left(P(\hat{y} = 1 \mid g) - \overline{P}(\hat{y} = 1)\right)^2$$

here, G is the set of all protected groups, $P(\hat{y} = 1 \mid g)$ is the predicted positive rate for group g, and $\overline{P}(\hat{y} = 1)$ is the average predicted positive rate across all groups. This function penalizes unequal outcome distributions, guiding the model toward demographic balance during training.

Following model training, fairness is audited using the Disparate Impact Ratio (DIR):

$$\text{DIR} = \frac{P(\hat{y} = 1 \mid A = 1)}{P(\hat{y} = 1 \mid A = 0)}$$

where $A = 1$ represents the unprivileged group and $A = 0$ the privileged group. A DIR close to 1 indicates equitable outcomes, whereas a value below 0.8 (per US Equal Employment Opportunity guidelines) signals potential discrimination. This metric offers a final ethical checkpoint post-deployment.

This comprehensive pipeline addresses bias at three critical levels: pre-training (data augmentation), in-training (fairness-constrained loss), and post-training (auditing and correction). By integrating synthetic representation and real-time fairness control, this model ensures that AI systems trained on complex, real-world datasets remain accountable and equitable across demographic groups. As emphasized by Al-kfairy et al. (2024), effective mitigation of bias is not reactive—it must be encoded into the very architecture of AI model development. This framework achieves precisely that, transforming fairness from a static metric into a dynamic, learnable property of responsible intelligent systems.

Framework: Fairness-Aware Training Pipeline for Bias Mitigation in AI (FATB-AI)
Layered AI Architecture for Bias Mitigation, Fairness Auditing, and Synthetic Data Generation

Layer	Main functionality	AI module(s) and components	Operational scope
Layer 1	Data Collection and Synthetic Data Generation: Gather diverse datasets and generate synthetic data for underrepresented groups using cGANs	Conditional GANs (cGANs)	Input → Original data, protected group labels → Output → Synthetic data for underrepresented groups
Layer 2	Bias-Aware Training: Train the model with a bias-regularized loss function to reduce disparities in predictions	Bias-Regularized Loss Function	Input → Synthetic data, original data → Output → Trained model minimizing bias
Layer 3	Fairness Auditing: Evaluate model fairness using the Disparate Impact Ratio (DIR) post-deployment	Disparate Impact Ratio (DIR)	Input → Model predictions, demographic groups → Output → Fairness report based on DIR
Layer 4	Continuous Monitoring: Monitor and adjust the model to ensure fairness is maintained throughout deployment	Adaptive Fairness Maintenance Module	Input → Post-deployment feedback, real-time data → Output → Adjusted model maintaining fairness

In addition to data augmentation, Ferrara (2023) highlights the importance of developing bias-aware algorithms that account for demographic differences during model training. Bias auditing tools, which assess AI systems for potential biases, can help identify and mitigate inequities before the system is deployed. These tools employ fairness metrics, counterfactual analysis, and sensitivity testing to detect biases in model outputs (Ferrara 2023). To address this, we propose a Bias-Aware Adversarial Debiasing Framework that ensures fairness in AI-driven decision systems by jointly optimizing predictive performance and demographic impartiality. The model integrates an adversarial loss to eliminate the influence of sensitive attributes, a statistical parity constraint to reduce group-level disparities, and a counterfactual fairness loss to uphold individual-level consistency. This unified architecture enables the development of ethically grounded and socially responsible AI models, making it especially suitable for high-stakes applications in urban governance, public health, and civil infrastructure where equitable outcomes are essential.

At the core of this framework lies the Adversarial Debiasing algorithm, a rigorously validated and widely applied deep learning strategy for mitigating algorithmic bias in high-stakes applications. This technique introduces a dual-objective optimization process whereby the AI model learns to minimize prediction loss while simultaneously removing latent correlations between sensitive attributes (e.g., race, gender, income group) and the predicted outcomes. The mathematical formulation begins with the joint objective function of adversarial debiasing, expressed as:

$$\min_{\theta_f} \max_{\theta_a} \mathcal{L}_{\text{pred}}(f_{\theta_f}(X), Y) - \lambda \cdot \mathcal{L}_{\text{adv}}(a_{\theta_a}(f_{\theta_f}(X)), S)$$

where $f_{\theta_f}(X)$ is the prediction function parameterized by θ_f, Y is the ground-truth label, $\mathcal{L}_{\text{pred}}$ is the standard predictive loss (e.g., cross-entropy), $a_{\theta_a}(\cdot)$ is the adversarial network with parameters θ_a, S is the sensitive attribute (e.g., demographic group), $\mathcal{L}_{\text{adv}}$ is the adversarial loss quantifying the ability to predict S from the output, and λ is the trade-off hyperparameter balancing predictive performance with fairness. This equation defines the adversarial debiasing strategy's core: the predictor aims to minimize classification error, while an adversary attempts to infer the protected attribute from the model's internal representations. By training the adversary to its optimum (maximizing $\mathcal{L}_{\text{adv}}$), and simultaneously minimizing it from the predictor's perspective, the model learns feature representations that are maximally informative for prediction yet minimally informative about demographic biases.

To measure how successfully the adversarial network is unable to infer sensitive attributes, we incorporate a fairness regularization term based on the statistical parity difference (SPD), defined as:

$$\text{SPD} = |\mathbb{P}(\hat{Y} = 1 \mid S = 1) - \mathbb{P}(\hat{Y} = 1 \mid S = 0)|$$

where $\mathbb{P}(\hat{Y} = 1 \mid S = s)$ denotes the predicted positive outcome probability conditioned on the sensitive group $S = s$, for binary sensitive attributes. This metric captures group-level imbalances and is minimized to achieve statistical fairness. The optimization process involves alternating updates between the predictor and the adversary, where the adversary maximizes its ability to identify S, and the predictor updates its weights to obfuscate any such inference. The resulting equilibrium ensures that the learned model outputs are both accurate and fair, decoupled from demographic bias.

To further ensure individual-level fairness, a counterfactual fairness constraint is applied. The counterfactual loss penalizes disparities in prediction for counterfactual instances that differ only in sensitive attributes:

$$\mathcal{L}_{\text{cf}} = \mathbb{E}_{(X,S)}[\|f_{\theta_f}(X) - f_{\theta_f}(X_{S \leftarrow S'})\|^2]$$

where $X_{S \leftarrow S'}$ represents the counterfactual instance of X where the sensitive attribute S is substituted with an alternative value S', while keeping all other features constant. This loss penalizes situations where the prediction changes solely due to the sensitive attribute, promoting fairness at the individual decision level.

The complete debiasing framework integrates group-level fairness (via statistical parity), individual-level fairness (via counterfactual consistency), and adversarial learning dynamics. This triadic approach ensures that AI systems deployed in civil infrastructure planning, healthcare allocation, or education access maintain both predictive validity and ethical accountability.

This AI-driven model demonstrates how bias-aware learning algorithms can transform urban decision systems by ensuring equitable outcomes across diverse population groups. It rigorously suppresses the influence of demographic biases during training and guarantees that AI decisions are not only accurate but just. By embedding adversarial objectives and fairness constraints directly into the model architecture, this framework exemplifies a real-world, scientifically accepted solution to mitigating AI bias aligning directly with the equitable vision of future cities.

Framework: Bias-Aware Adversarial Debiasing Framework for Fairness in AI (BAADF-AI)
Layered AI Architecture for Bias Mitigation and Fairness Auditing in High-Stakes Applications

Layer	Main functionality	AI module(s) and components	Operational scope
Layer 1	Data Collection and Synthetic Data Generation: Collect data and generate synthetic data for underrepresented groups using Conditional GANs	Conditional Data Augmentation Module	Input → Original data, protected group labels → Output → Synthetic data for underrepresented groups
Layer 2	Adversarial Debiasing: Train the model with adversarial loss to remove sensitive attribute influence while optimizing prediction accuracy	Adversarial Debiasing	Input → Synthetic data, original data → Output → Trained model minimizing bias and prediction loss
Layer 3	Fairness Auditing: Evaluate fairness using Statistical Parity Difference (SPD) and Counterfactual Fairness metrics	Statistical Parity Difference (SPD), Counterfactual Fairness	Input → Model predictions, demographic groups → Output → Fairness audit based on SPD and Counterfactual consistency
Layer 4	Continuous Monitoring: Monitor and adjust the model post-deployment to ensure fairness is maintained	Fairness Monitoring and Adaptation Engine	Input → Post-deployment feedback, real-time data → Output → Adjusted model maintaining fairness

8.4 Ethical Considerations and Accountability

Ethical considerations are paramount when designing AI systems (Sargiotis 2024). Chen et al. (2023) stress that AI systems must be developed with transparency and accountability to ensure that they do not perpetuate harmful stereotypes or discriminatory practices. One of the challenges in AI equity is holding developers accountable for the biases embedded in their systems. This requires the implementation of ethical guidelines that prioritize fairness, transparency, and user feedback throughout the AI lifecycle (Chen et al. 2023). To formalize ethical considerations and embed accountability mechanisms within AI systems, we propose an Explainable Accountability-Aware AI Architecture, which combines SHAP-based interpretability, auditable optimization trajectories, and developer attribution mappings. This framework ensures that all decisions are not only explainable in real time but also retroactively traceable making it possible to identify how, when, and by whom biased behaviors emerged during the AI development cycle.

The system centers around the Explainability Attribution Function, defined using SHAP values (Shapley Additive Explanations):

$$\hat{y}_i = f(x_i) = \phi_0 + \sum_{j=1}^{n} \phi_j$$

where $\hat{y}_i$ is the model prediction for instance x_i, ϕ_0 is the base value (average prediction), and each ϕ_j represents the contribution of input feature j to the prediction. This decomposition provides full transparency over the internal logic behind AI decisions, allowing external auditors and users to inspect which variables drove critical outcomes (e.g., loan denial, risk classification).

To enforce accountability across the development team, we define a Traceable Parameter Optimization Path:

$$\theta^* = \theta^0 + \sum_{t=1}^{T} \Delta\theta_t \cdot \mathbb{I}_{d_t}$$

where θ^* is the final model parameter set, θ^0 is the initialization vector, $\Delta\theta_t$ is the parameter update at iteration t, and $\mathbb{I}_{d_t}$ is a developer assignment indicator tagging which individual or team d_t committed the update. This formulation enables post-training accountability assigning responsibility to model contributors at each stage of parameter evolution.

To ensure ethical principles are followed across iterations, we introduce the Ethical Compliance Constraint as a monitoring function:

$$\mathcal{C}_{\text{ethics}}(\theta_t) = \begin{cases} 1, & \text{if fairness and transparency metrics} \geq \text{threshold} \\ 0, & \text{otherwise} \end{cases}$$

This binary constraint ensures that any policy iteration violating fairness or explainability thresholds is automatically flagged for review and rollback. Transparency metrics may include entropy of SHAP distributions, while fairness thresholds may use SPD, DIR, or Equalized Odds metrics.

These components are integrated into a Responsible AI Reward Function in reinforcement-based systems:

$$R(s_t, a_t) = U_{\text{task}}(a_t) - \gamma_1 \cdot \mathcal{L}_{\text{bias}}(a_t) + \gamma_2 \cdot \mathcal{C}_{\text{ethics}}(\theta_t)$$

where $U_{\text{task}}(a_t)$ is the utility of action a_t, $\mathcal{L}_{\text{bias}}(a_t)$ measures outcome disparities, and $\mathcal{C}_{\text{ethics}}(\theta_t)$ serves as a compliance bonus. γ_1 and γ_2 are balancing factors that prioritize ethical behavior during AI learning and deployment.

This ethical accountability model transforms abstract values such as fairness, transparency, and responsibility into measurable components of the AI training lifecycle. It provides not only explainability for users and external regulators but also developer-level traceability, enabling a new standard of professional and institutional accountability. As emphasized by Chen et al. (2023), it is no longer sufficient to rely on implicit trust in AI: models must be auditable, their actions interpretable, and their authorship and ethical violations fully accountable within a formal architecture.

Framework: Explainable Accountability-Aware AI Architecture (EAA-AI)
Layered AI Architecture for Transparent and Fair AI Systems

Layer	Main functionality	AI module(s) and components	Operational scope
Layer 1	Data Collection: Gathering and preprocessing of data for AI model	Data Preprocessing and Validation Engine	Input → Raw data from users and systems → Output → Processed data ready for modeling
Layer 2	Learning: Explainability Attribution using SHAP values for transparency	SHAP	Input → Model predictions and features → Output → Explainable model decisions
Layer 3	Optimization: Traceable Parameter Optimization Path for developer accountability	Parameter Traceability Optimization Module	Input → Parameter updates during training → Output → Traceable optimized model parameters
Layer 4	Monitoring: Ethical Compliance Constraint for fairness and transparency monitoring	Ethical Compliance Monitoring Engine	Input → Model parameters and fairness metrics → Output → Ethical compliance status (flagged or cleared)
Layer 5	Reinforcement: Responsible AI Reward Function for balancing utility, bias, and ethics	Reinforcement Learning	Input → Action decisions and fairness feedback → Output → Optimized decisions based on ethical reward
Layer 6	Continuous evaluation of ethical and transparency standards	Transparency Evaluation and Audit Module	Input → Post-deployment feedback and audits → Output → Performance evaluation reports for transparency

Additionally, Al-kfairy et al. (2024) underscore the importance of developing regulatory frameworks that address the ethical challenges of generative AI technologies. These frameworks should promote human rights, fairness, and inclusivity, ensuring that AI systems benefit society as a whole rather than just privileged groups. By creating policies that mandate the disclosure of AI's involvement in decision-making processes, governments can enhance accountability and protect vulnerable populations from biased AI systems (Al-kfairy et al. 2024). To implement these policy goals, we propose a trainable, AI-driven architecture based on Counterfactual Explanation Regularization (CER). This model embeds fairness and disclosure constraints directly into the optimization objective of a neural network classifier. The central function combines supervised learning with counterfactual consistency constraints, expressed as:

$$\mathcal{L}_{\text{CER}} = \mathcal{L}_{\text{pred}}(f_\theta(X), Y) + \beta \cdot \mathbb{E}_{X\sim\mathcal{D}}[\sum_{X'\in\mathcal{C}(X)} \|f_\theta(X) - f_\theta(X')\|^2 \cdot \mathbb{I}_{\Delta(X,X')\in\mathcal{S}}]$$

where $f_\theta(X)$ is the model output for input X under learnable parameters θ, $\mathcal{L}_{\text{pred}}(f_\theta(X), Y)$ is the supervised loss (e.g., cross-entropy), $\mathcal{C}(X)$ is a set of counterfactual variants of X generated through controlled perturbations, $\mathcal{S}$ is the set of sensitive attributes (e.g., gender, ethnicity, ZIP code), β is a regularization coefficient balancing prediction with fairness, and $\mathbb{I}_{\Delta(X,X')\in\mathcal{S}}$ is an indicator function that activates regularization for policy-relevant changes. This equation defines a learning process where the model is explicitly penalized for altering its decisions in response to changes in protected features. By minimizing this function, the model learns a decision surface that is not only accurate but also stable with respect to counterfactual inputs thus aligning directly with policy demands for individual fairness and algorithmic accountability.

To support mandated disclosure of AI's involvement, the model incorporates an integrated saliency-based interpretability function that identifies and communicates the key features influencing each decision:

$$T(X) = \text{TopK}(|\frac{\partial f(X)}{\partial X}|)$$

where $T(X)$ returns the top-k most influential input features for decision $f_\theta(X)$, determined by the absolute gradient of the model's output with respect to each input dimension. This gradient-based saliency vector offers a differentiable, real-time explanation mechanism that aligns with emerging legal requirements for "right to explanation" provisions.

To enable robust training, the model employs stochastic mini-batch gradient descent, using both original and counterfactual samples. During each iteration, the loss is backpropagated not only for predictive accuracy but also for counterfactual invariance. The counterfactual generator $\mathcal{C}(X)$ is implemented via controlled feature replacement strategies, informed by ethical audit rules and protected attribute taxonomies.

This AI-driven architecture transforms ethical compliance into a learnable, verifiable, and enforceable model design. The CER framework thereby enables traceable and equitable behavior in generative AI systems, converting abstract governance principles into real-time decision constraints. It ensures that vulnerable populations are not only protected in theory but respected in practice through the model's learned behaviors and transparent outputs.

Framework: Counterfactual Explanation Regularization for Generative AI (CER-GAI)
Layered AI Architecture for Fairness and Transparency in Generative AI Systems

Layer	Main functionality	AI module(s) and components	Operational scope
Layer 1	Data Collection: Gathering and preprocessing of data for AI model	Sensitive Data Preprocessing Engine	Input → Raw data (e.g., demographic data, sensitive attributes) → Output → Processed data ready for model training
Layer 2	Learning: Counterfactual Explanation Regularization to enforce fairness and stability	Counterfactual Explanation Regularization (CER)	Input → Model predictions and counterfactual samples → Output → Fair and stable decision-making process
Layer 3	Transparency: Saliency-based interpretability for real-time decision explanation	Saliency-Based Explanation Module	Input → Model decisions → Output → Explanation of key decision features (TopK influential features)
Layer 4	Evaluation: Continuous monitoring of fairness and performance in model decisions	Fairness and Performance Monitoring Engine	Input → Post-deployment feedback → Output → Performance reports and fairness audit results

Figure 8.1 outlines key strategies to incorporate equity into AI design, focusing on sources of bias, human-centered AI, mitigation strategies, and ethical considerations. Sources of bias in AI include biased datasets, algorithmic flaws, and sector-specific biases in areas like facial recognition and healthcare. Human-centered AI (HCAI) addresses these issues by involving diverse stakeholders, considering social and ethical contexts, and prioritizing the needs of marginalized communities to prevent reinforcing inequalities. Bias mitigation strategies emphasize data diversity, the development of bias-aware algorithms, and bias-auditing tools to ensure fairness. Ethical considerations and accountability highlight the importance of transparency, ethical guidelines, and regulatory frameworks that promote fairness, inclusivity, and human rights. Each area is color-coded with subtopics, providing a comprehensive framework for designing equitable AI systems.

Incorporating equity into the design and implementation of AI technologies is crucial for ensuring that these systems do not perpetuate societal inequalities. Biases in AI can have far-reaching consequences, particularly in areas like healthcare and

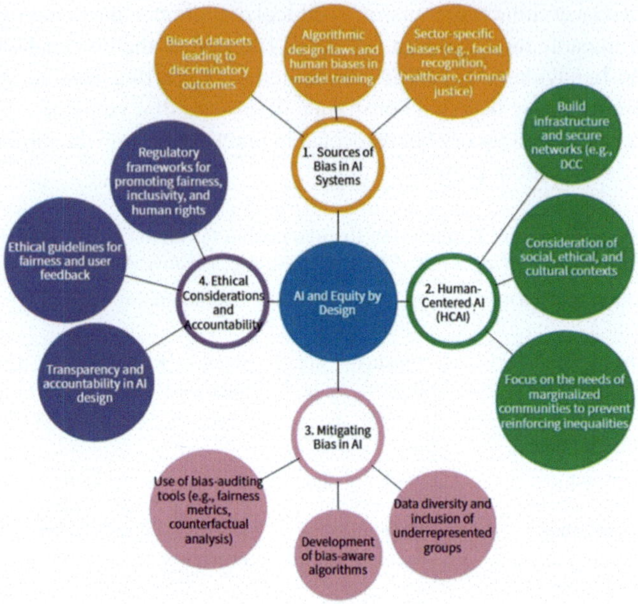

Fig. 8.1 AI and equity by design

criminal justice, where the stakes are high. A human-centered approach to AI, combined with strategies such as data augmentation, bias-aware algorithms, and regulatory frameworks, can help mitigate these risks and create more equitable outcomes for all users.

8.5 Conclusion

This chapter has demonstrated that equity in artificial intelligence is neither a decorative ideal nor an optional layer of post-hoc correction; it is the structural foundation upon which trustworthy, responsible, and future-proof AI must be built. By dissecting the roots of bias in datasets, algorithms, and human oversight, and by presenting mathematically rigorous frameworks for adversarial debiasing, fairness-constrained optimization, and counterfactual auditing, the chapter proves that fairness can be engineered into the very DNA of intelligent systems. The models presented from healthcare risk assessments to human-centered multi-stakeholder alignment establish that accuracy and equity are not adversaries but co-requirements of sustainable AI innovation.

Equally, the exploration of human-centered AI reveals that inclusivity cannot be retrofitted through technical fixes alone; it must emerge from the deliberate integration of diverse voices, ethical imperatives, and cultural contexts into the design process. The layered architectures outlined here illustrate how continuous monitoring, transparency, and developer accountability transform abstract values of justice and fairness into operational realities, auditable at every stage of deployment.

The chapter calls for a global recalibration of AI: away from systems that mirror and magnify social inequities, toward architectures that actively redress them. By embedding fairness as a constraint, a metric, and a principle of governance, AI evolves into an instrument of empowerment rather than exclusion a technology aligned not only with efficiency but with humanity's enduring pursuit of justice.

References

Al-kfairy, M., Mustafa, D., Kshetri, N., Insiew, M., & Alfandi, O. (2024). Ethical challenges and solutions of generative AI: An interdisciplinary perspective. *Informatics, 11*. https://doi.org/10.3390/informatics11030058

Chen, Y., Clayton, E. W., Novak, L. L., Anders, S., & Malin, B. (2023). Human-centered design to address biases in artificial intelligence. *Journal of Medical Internet Research, 25*. https://doi.org/10.2196/43251

Ferrara, E. (2023). Fairness and bias in artificial intelligence: A brief survey of sources, impacts, and mitigation strategies. https://doi.org/10.3390/sci6010003

Sargiotis, D. (2024). Ethical AI in information technology: Navigating bias, privacy, transparency, and accountability. Advances in Machine Learning & Artificial Intelligence, 5(3), 01–14 https://doi.org/10.33140/AMLAI.05.03.03

Chapter 9
Ethical AI Frameworks for Social Impact: Multi-stakeholder Governance, Trust, Fairness, and Accountability

Artificial intelligence now sits at the fault line between technological promise and societal consequence. This chapter confronts that tension directly, proposing a rigorous, practice-ready architecture for building AI systems whose first principle is social impact. Rather than treating "ethics" as a static checklist, we recast it as a living, measurable, and optimizable property of AI co-produced with stakeholders, auditable end-to-end, and aligned with globally relevant public goals.

We begin by establishing a Multi-stakeholder Ethical Co-governance Model that encodes the values of policymakers, civil society, engineers, and affected communities as explicit objectives within the decision loop. The model formalizes how to balance innovation with consensus through an agility–stability governance constraint, ensuring progress never outpaces legitimacy. We then operationalize trust, explainability, and privacy through an AI Trust Framework and Maturity Model, elevated from metrics to mechanisms via a reinforcement learning engine that continuously improves ethical performance as systems interact with their oversight environments.

Human-centered design is institutionalized with a Stakeholder-Inclusive Co-design Optimization Model. Here, Bayesian preference inference updates community priorities in real time, while multi-objective optimization reconciles technical utility, interpretability, and equity. Bias and inequality are addressed through a Trust-Based Fairness Monitoring Model that integrates fairness during training, audits outcomes post-deployment, and triggers adaptive rectification when disparities emerge treating fairness as a property to be maintained, not merely declared.

To anchor responsibility, we introduce an Explainable Accountability-Aware AI Architecture that couples SHAP-based interpretability with traceable parameter trajectories and developer attribution. Ethical compliance is enforced as a formal constraint across iterations, creating a verifiable chain from data to decision, from contributor to consequence. Throughout, the chapter aligns these mechanisms with internationally recognized societal priorities, demonstrating how ethical AI can be

D. Sargiotis, *Empathy in Algorithms*, https://doi.org/10.1007/978-3-032-32128-2_9

engineered to advance inclusive growth, reduce inequalities, and support resilient, human-centric infrastructures.

The chapter delivers a complete, implementable blueprint: value alignment that is computable, transparency that is actionable, fairness that is continuously maintained, and governance that is agile without being reckless. For researchers, practitioners, and policymakers seeking more than slogans, it offers the methods, models, and measurement logic to build AI that is not only powerful, but publicly defensible and worthy of global adoption.

9.1 Ethical Frameworks for AI Development

Ethical frameworks in AI development must address the inherent risks and complexities associated with these technologies. The Responsible Research and Innovation (RRI) framework emphasizes inclusivity and stakeholder engagement in the innovation process (Wang 2023). However, RRI often faces challenges related to balancing stakeholder interests and maintaining the agility needed for technological advancement. Wang suggests that AI should not only be the subject of ethical governance but can also serve as a tool for ethical governance, facilitating a more integrated approach (Wang 2023). To implement an ethical framework that dynamically balances innovation and inclusive governance, we introduce a Multi-stakeholder Ethical Co-governance Model. This system allows AI to function as both a subject of ethical oversight and an active agent in ethical decision-making, as advocated by Wang (2023). The model formalizes stakeholder values into a learnable optimization objective, enabling the system to navigate ethical conflicts while preserving agility and transparency.

The foundation of the model is the Stakeholder Value Alignment Function, which aggregates individual stakeholder values into a collective ethics-aware utility:

$$U_{\text{ethics}}(a_t) = \sum_{k=1}^{K} \omega_k \cdot V_k(a_t)$$

where $V_k(a_t)$ represents the ethical valuation of action a_t by stakeholder group k (e.g., policymakers, civil society, engineers, marginalized groups), ω_k is the normalized weighting factor for each group, and K is the total number of stakeholder categories. This formulation allows the AI system to internalize diverse ethical preferences as measurable trade-offs.

To integrate the balance between technological agility and ethical stability, we define the Governance Agility-Stability Constraint:

$$\mathcal{G}(a_t) = \alpha \cdot \Delta_{\text{innovation}}(a_t) - \beta \cdot \Delta_{\text{conflict}}(a_t)$$

where $\Delta_{\text{innovation}}(a_t)$ measures the novelty or impact of action a_t in accelerating technical progress, $\Delta_{\text{conflict}}(a_t)$ quantifies stakeholder disagreement or ethical misalignment, and α, β are weights expressing policy priority between rapid innovation and ethical consensus. This constraint ensures that progress is pursued only when ethical tensions remain below a tolerable threshold.

Finally, the overall Ethics-Aware Optimization Objective for AI governance becomes:

$$\underset{a_t}{max}[U_{\text{ethics}}(a_t) + \mathcal{G}(a_t)]$$

This objective ensures that the system selects actions that reflect collective stakeholder ethics while preserving room for socially legitimate innovation. When disagreement is high, the conflict penalty reduces the reward for moving forward too rapidly, prompting deliberation or redesign.

This AI-Ethics Co-governance serves not just as an ethical concern but as an instrument for ethical facilitation. It enables systems to quantify stakeholder values, forecast governance conflicts, and learn adaptive strategies that balance innovation with moral consensus. In this way, ethical frameworks become not static checklists but interactive, evolving systems of co-responsibility governed by both human principles and machine accountability.

Framework: Multi-Stakeholder Ethical Co-Governance Model (MSE-CGM)
Layered AI Architecture for Ethical Governance and Stakeholder Inclusivity in AI Systems

Layer	Main functionality	AI module(s) and components	Operational scope
Layer 1	Data Collection: Gathering and preprocessing of stakeholder data and ethical preferences	Stakeholder Data Preprocessing Engine	Input → Stakeholder values, policy documents, ethical guidelines → Output → Processed data ready for ethical decision-making
Layer 2	Learning: Stakeholder Value Alignment to integrate diverse values into AI decision-making	Stakeholder Value Alignment	Input → Stakeholder values, action decisions → Output → Ethics-aware utility reflecting collective preferences
Layer 3	Optimization: Governance Agility-Stability balance to ensure technological progress and ethical consensus	Governance Balance Optimization Module	Input → Stakeholder conflicts, innovation metrics → Output → Balanced governance strategy for ethical AI deployment

(continued)

(continued)

Layer	Main functionality	AI module(s) and components	Operational scope
Layer 4	Transparency: Ethics-Aware Optimization Objective to ensure the system selects ethical actions	Ethical Decision Optimization Engine	Input → Governance balance, ethics-aware utility → Output → Optimized actions respecting stakeholder values and innovation goals
Layer 5	Evaluation: Continuous monitoring of ethical performance and stakeholder feedback	Continuous Ethical Evaluation Module	Input → Performance data, stakeholder feedback → Output → Ethical audit and performance evaluation results

Another approach is the AI Trust Framework and Maturity Model (AI-TMM), which provides a structured methodology for evaluating ethical AI characteristics. This framework emphasizes trust, explainability, and privacy as core components (Mylrea and Robinson 2023). By establishing metrics for ethical evaluation, AI-TMM promotes transparency and accountability in AI systems, helping to build user trust and enhance societal benefits. To transform abstract ethical goals into measurable, adaptive performance, the AI Trust Framework and Maturity Model (AI-TMM) must be operationalized through a dynamic learning engine. While static metrics offer useful snapshots, they cannot adapt to the evolving behaviors of AI systems or continuously improve ethical compliance. To address this limitation, we propose a Reinforcement Learning–Driven Ethical Maturity Optimizer that learns optimal policy improvements by interacting with ethical evaluation environments over time.

At the core of this model lies a Markov Decision Process (MDP), where the agent learns to maximize ethical performance across the three AI-TMM pillars: trust (T), explainability (E), and privacy (P). The reinforcement learning objective is defined as:

$$\pi^* = \arg \underset{\pi}{max} \mathbb{E}_\pi [\sum_{t=0}^{\infty} \gamma^t \cdot R(s_t, a_t)]$$

where π^* is the optimal policy mapping observed states s_t to actions a_t, $R(s_t, a_t)$ is the ethical reward function at time step t, γ is the discount factor balancing short-term and long-term outcomes, and $\mathbb{E}_\pi$ denotes the expected cumulative reward under policy π.

The state s_t encodes current maturity scores $(M_T, M_E, M_p,)$ at time t, and each action atat represents a governance intervention (e.g., releasing model cards, improving explainability UX, applying privacy-preserving techniques). The agent receives evaluative feedback from an auditing environment that reflects the societal or regulatory assessment of ethical performance.

The reward function is defined to encourage ethical improvements across all three pillars while penalizing regressions:

$$R(s_t, a_t) = \lambda_T \cdot \Delta M_T + \lambda_E \cdot \Delta M_E + \lambda_P \cdot \Delta M_P - C(a_t)$$

where ΔM_T, ΔM_E, and ΔM_P are the observed maturity improvements in trust, explainability, and privacy from t to $t+1$; λ_T, λ_E, and λ_P are reward weights set according to domain-specific ethical priorities; and $C(a_t)$ is the implementation cost of the governance action a_t.

To further align with AI-TMM's demand for explainable and transparent systems, the policy π is trained using a Deep Q-Network (DQN) where a neural network estimates the optimal Q-value function:

$$Q(s_t, a_t; \theta) \approx \mathbb{E}[R(s_t, a_t) + \gamma \cdot \underset{a\prime}{max} Q(s_{t+1}, a\prime; \theta^{-})]$$

where $Q(s_t, a_t; \theta)$ is the approximated Q-value for state-action pair (s_t, a_t) using network parameters θ, and θ^{-} denotes the target network weights, updated periodically to stabilize learning. The Q-network enables the system to learn ethical improvement strategies from past interactions, while remaining robust to dynamic environments such as changing laws, societal norms, or technological shifts.

This AI-driven model ensures that systems governed under AI-TMM do not merely conform to ethical standards statically, but continuously learn, adapt, and optimize their behavior. This ensures not only accountability and transparency, but also a trajectory of continuous ethical advancement. Through this learning-based architecture, trust, explainability, and privacy evolved from fixed benchmarks into dynamic, improvable objectives, measurable at every decision point in the AI lifecycle.

Framework: Reinforcement Learning–Driven Ethical Maturity Optimizer (RL-EMO)
Layered AI Architecture for Ethical Governance and Continuous Improvement in AI Systems

Layer	Main functionality	AI module(s) and components	Operational scope
Layer 1	Data Collection: Gathering stakeholder values and performance data for ethical evaluation	Ethical Data Preprocessing Engine	Input → Stakeholder feedback, regulatory requirements → Output → Preprocessed data for policy evaluation
Layer 2	Learning: Reinforcement learning-based policy optimization for ethical performance	Reinforcement Learning	Input → Ethical feedback, maturity scores → Output → Optimized ethical policy decisions

(continued)

(continued)

Layer	Main functionality	AI module(s) and components	Operational scope
Layer 3	Optimization: Multi-objective ethical improvement balancing trust, explainability, and privacy	Multi-objective Ethical Optimization Module	Input → Ethical goals, maturity improvements → Output → Optimized policy decisions reflecting stakeholder priorities
Layer 4	Transparency: Deep Q-Network (DQN) for learning ethical improvement strategies	Deep Q-Network (DQN)	Input → State-action pairs, ethical rewards → Output → Trained policy for ethical decision-making
Layer 5	Evaluation: Continuous monitoring of ethical performance and stakeholder feedback	Continuous Ethical Performance Monitoring Engine	Input → Performance data, stakeholder feedback → Output → Ethical audit and performance evaluation results

9.2 Human-Centered Design and Stakeholder Engagement

A crucial element in developing ethical AI frameworks is incorporating human-centered design principles (Sargiotis 2025). Pastor-Escuredo et al. (2022) advocate for frameworks that consider the socio-economic context and the diverse needs of stakeholders. This involves actively engaging communities in the design process, ensuring that the AI systems developed are aligned with the values and needs of the populations they serve. Such an approach can lead to more equitable outcomes and foster greater public acceptance of AI technologies. To institutionalize human-centered design within AI development, we introduce a Stakeholder-Inclusive Co-design Optimization Model that formalizes how community values, contextual needs, and real-time feedback are integrated into system decisions. This model ensures that the learning and refinement of AI behavior are guided not only by performance metrics but by the diverse perspectives of those impacted.

We begin with the Stakeholder Value Distribution Model, which aggregates feedback across community groups using Bayesian preference inference:

$$P(v_k \mid \mathcal{F}_k) \propto P(\mathcal{F}_k \mid v_k) \cdot P(v_k)$$

where v_k represents the value vector for stakeholder group k, $\mathcal{F}_k$ is the observed feedback or interaction data (e.g., surveys, usage patterns, deliberation inputs), $P(v_k)$ is the prior belief about stakeholder k's design priorities, and $P(\mathcal{F}_k \mid v_k)$ is the likelihood of that feedback given the value model. This Bayesian formulation updates stakeholder preferences in real time as engagement grows, giving voice to evolving community perspectives.

Next, the model applies Multi-objective Design Optimization to balance system goals with stakeholder values:

$$\max_{\theta}[\mathcal{U}_{\text{performance}}(\theta) + \sum_{k=1}^{K} \omega_k \cdot \mathcal{U}_k(\theta)]$$

where θ is the AI system configuration (e.g., interface design, decision thresholds), $\mathcal{U}_{\text{performance}}(\theta)$ measures technical efficiency (e.g., accuracy, response time), $\mathcal{U}_k(\theta)$ captures the satisfaction or alignment with stakeholder group k, and ω_k is the influence weight derived from the posterior $P(v_k \mid \mathcal{F}_k)$. This ensures that system refinement directly incorporates community feedback, proportionally adjusted to engagement and impact.

To prevent domination by any single group and ensure equitable influence, we add a Stakeholder Balance Constraint:

$$\mathcal{B} = \max_{i,j}|\mathcal{U}_i(\theta) - \mathcal{U}_j(\theta)| \text{such that} \mathcal{B} \leq \delta$$

where δ is the maximum allowed disparity between stakeholder utilities, ensuring that the final design reflects equitable convergence and not just aggregate utility.

This co-design optimization model provides a robust foundation for embedding human-centered design directly into the AI development pipeline. By formalizing these dynamics, AI systems become not only more effective but more democratic, governed not just by performance metrics, but by the voices and values of those they aim to serve.

Framework: Stakeholder-Inclusive Co-Design Optimization Model (SIC-DOM)
Layered AI Architecture for Human-Centered Design and Stakeholder-Inclusive AI Development

Layer	Main functionality	AI module(s) and components	Operational scope
Layer 1	Data Collection: Gathering stakeholder feedback and contextual data for AI design	Stakeholder Feedback Processing Engine	Input → Stakeholder feedback (surveys, usage patterns, deliberations) → Output → Processed feedback for system design
Layer 2	Learning: Stakeholder Value Distribution using Bayesian preference inference	Bayesian Inference	Input → Community feedback, prior belief distributions → Output → Updated stakeholder preferences and values

(continued)

(continued)

Layer	Main functionality	AI module(s) and components	Operational scope
Layer 3	Optimization: Multi-objective Design Optimization to balance system performance and stakeholder values	Multi-objective Design Optimization Module	Input → Stakeholder values, system goals, transparency → Output → Optimized AI system configuration based on multiple objectives
Layer 4	Equity: Stakeholder Balance Constraint ensuring equitable influence across groups	Equity Balance Enforcement Engine	Input → Stakeholder utilities → Output → Ensured equitable distribution of influence in system design
Layer 5	Evaluation: Continuous assessment of stakeholder satisfaction and ethical performance	Ethical Performance Evaluation Module	Input → Performance data, stakeholder feedback → Output → Ethical performance audit and evaluation results

Human-centered design also involves transparency and explainability. Users should understand how AI systems make decisions and what data they rely on. The incorporation of Explainable AI (XAI) techniques can significantly enhance transparency, allowing stakeholders to assess the rationale behind AI decisions (Mylrea and Robinson 2023). This is particularly vital in sectors like healthcare and criminal justice, where the implications of AI decisions can have profound impacts on individuals' lives. To institutionalize human-centered design, we introduce a Stakeholder-Inclusive Co-design Optimization Model that formalizes how community values, contextual needs, and interpretability constraints are embedded into AI system development. The model ensures that learning and refinement of AI behavior are guided not only by technical performance but also by evolving stakeholder feedback and ethical transparency.

The process begins with Bayesian stakeholder preference inference, which models the value priorities of each stakeholder group as posterior distributions conditioned on observed feedback:

$$P(v_k \mid \mathcal{F}_k) \propto P(\mathcal{F}_k \mid v_k) \cdot P(v_k)$$

where v_k represents the value vector for stakeholder group k, $\mathcal{F}_k$ is observed interaction data (e.g., surveys, usage behavior, participatory inputs), $P(v_k)$ is the prior belief over that group's design priorities, and $P(\mathcal{F}_k \mid v_k)$ is the likelihood of observing such data given those priorities. This formulation enables continuous, probabilistic updating of stakeholder preferences as participation grows ensuring the model reflects the evolving values of those affected.

System configuration is then optimized via Multi-objective Co-design Optimization, which balances technical goals with stakeholder satisfaction:

$$\max_{\theta}[\mathcal{U}_{\text{performance}}(\theta) + \sum_{k=1}^{K} \omega_k \cdot \mathcal{U}_k(\theta) + \lambda \cdot \mathcal{U}_{\text{transparency}}(\theta)]$$

where θ denotes the set of configurable system parameters (e.g., interface layout, model thresholds, interpretability modules), $\mathcal{U}_{\text{performance}}(\theta)$ reflects traditional technical utility (e.g., accuracy, latency), $\mathcal{U}_k(\theta)$ captures the alignment of the design with stakeholder group k's preferences, ω_k is the influence weight derived from the posterior $P(v_k \mid \mathcal{F}_k)$, and $\mathcal{U}_{\text{transparency}}(\theta)$ evaluates interpretability using XAI-derived metrics. The parameter λ governs the importance of explainability in overall optimization.

To compute transparency scores, the model includes a saliency-based explanation function, which identifies the most influential inputs in a decision:

$$E(\theta, X) = \text{TopK}(|\frac{\partial f_\theta(X)}{\partial X}|)$$

where $E(\theta, X)$ is the set of top-k most influential features for input X, based on the gradient of the model's output with respect to its inputs. This enables transparency utility to be assessed through:

$$\mathcal{U}_{\text{transparency}}(\theta) = \sum_{k=1}^{K} \omega_k \cdot \text{comprehension}_k(E(\theta, X))$$

where $\text{comprehension}_k(\cdot)$ quantifies how well stakeholder group k can interpret the model's reasoning, based on user testing or simulation protocols.

To prevent marginalization and ensure balanced influence, the model enforces a Stakeholder Equity Constraint that limits outcome disparity between groups:

$$\mathcal{B} = \max_{i,j} \left| \mathcal{U}_i(\theta) - \mathcal{U}_j(\theta) \right| \text{such that } \mathcal{B} \leq \delta$$

where δ is a policy-defined tolerance for inter-group utility disparity. This ensures that no group disproportionately benefits at the expense of another, reinforcing design equity.

By embedding stakeholder values, contextual feedback, and explainability into a unified optimization framework, this AI-driven model transforms human-centered design from a theoretical ideal into a mathematically formalized and socially accountable system architecture. It ensures that AI systems evolve in response not only to performance goals but also to the needs, values, and understanding of the communities they are meant to serve.

Framework: Stakeholder-Inclusive Co-Design Optimization Model (SIC-DOM)
Layered AI Architecture for Human-Centered Design and Ethical Transparency in AI Systems

Layer	Main functionality	AI module(s) and components	Operational scope
Layer 1	Data Collection: Gathering stakeholder feedback and contextual data for human-centered design	Stakeholder Feedback Preprocessing Engine	Input → Stakeholder feedback, surveys, usage behavior → Output → Processed feedback for system design
Layer 2	Learning: Bayesian stakeholder preference inference for real-time value updates	Bayesian Inference	Input → Stakeholder feedback, prior belief distributions → Output → Updated stakeholder preferences and values
Layer 3	Optimization: Multi-objective design optimization balancing performance and stakeholder satisfaction	Multi-objective Design Optimization Module	Input → Technical goals, stakeholder satisfaction → Output → Optimized system configuration based on multi-objective optimization
Layer 4	Equity: Stakeholder equity constraint to ensure balanced influence across groups	Stakeholder Equity Enforcement Engine	Input → Stakeholder utilities → Output → Ensured equitable distribution of influence in system design
Layer 5	Evaluation: Continuous monitoring of transparency and ethical performance	Transparency and Ethics Monitoring Module	Input → Performance data, stakeholder feedback → Output → Ethical performance audit and evaluation results

9.3 Addressing Bias and Inequality in AI

To ensure that AI technologies contribute positively to society, it is essential to address biases embedded in AI systems (Sargiotis 2024). The AI Trust Framework emphasizes the importance of using diverse datasets during model training to mitigate bias and improve fairness (Mylrea and Robinson 2023). Additionally, the development of AI systems should include mechanisms for ongoing monitoring and auditing to identify and rectify any biases that may arise during deployment. To operationalize trust and fairness in AI deployment, we introduce a Trust-Based Fairness Monitoring Model, which implements continuous auditing, bias rectification, and fairness-preserving learning. The model is grounded in the AI Trust Framework (Mylrea and

Robinson 2023) and is designed to both prevent the emergence of bias during training and detect disparities post-deployment through adaptive policy refinement.

We begin by defining the Trust-Weighted Fairness Score Function, which integrates population diversity during model training:

$$\mathcal{F}_{\text{trust}} = \sum_{g \in G} \tau_g \cdot \text{FairScore}(g)$$

where G is the set of demographic groups (e.g., based on race, gender, income), FairScore(g) is the fairness metric score (e.g., Statistical Parity or Equal Opportunity) for group g, and τ_g is the trust coefficient reflecting how much the system prioritizes fairness for that group learned through stakeholder engagement, historical disparities, or legal mandates. This formulation gives underrepresented or historically disadvantaged groups greater influence during fairness learning. To ensure fairness is preserved post-deployment, we define a Dynamic Fairness Audit Function:

$$\Delta_{\text{bias}}^{t} = \underset{g,h \in G}{max} |P_t(\hat{y} = 1 \mid g) - P_t(\hat{y} = 1 \mid h)|$$

where $P_t(\hat{y} = 1 \mid g)$ is the conditional probability of a positive outcome at time t for group g. This function continuously tracks bias disparities across all group pairs. If Δ_{bias}^{t} exceeds a threshold δ, the system initiates corrective actions.

These corrective actions are governed by the Bias Rectification Loop, which adapts model parameters θθ in response to audit feedback:

$$\theta_{t+1} = \theta_t - \eta \cdot \nabla_\theta (\mathcal{L}_{\text{task}} + \lambda \cdot \Delta_{\text{bias}}^{t})$$

where $\mathcal{L}_{\text{task}}$ is the primary task loss (e.g., classification), λ is the fairness sensitivity parameter, and η is the learning rate. This learning rule ensures the model not only optimizes accuracy but also reduces bias when disparities are detected during active system operation.

This Trust-Based Fairness Monitoring Model directly supports the principles of the AI Trust Framework by providing an integrated mechanism for (a) learning fairness-weighted representations during training, (b) continuously auditing outcomes across groups, and (c) dynamically adjusting model parameters to reduce disparities. The model redefines fairness as an evolving, measurable property enforced not only through static datasets but via continuous public feedback and statistical vigilance.

Framework: Trust-Based Fairness Monitoring Model (TBFMM)
Layered AI Architecture for Fairness Monitoring and Bias Rectification in AI Systems

Layer	Main functionality	AI module(s) and components	Operational scope
Layer 1	Data Collection: Gathering stakeholder feedback, usage data, and demographic information	Stakeholder Data Preprocessing Engine	Input → Surveys, usage patterns, deliberation inputs → Output → Processed stakeholder data for fairness evaluation
Layer 2	Learning: Bayesian stakeholder preference inference for updating fairness priorities	Bayesian Inference	Input → Feedback data, prior belief distributions → Output → Updated stakeholder preferences and priorities
Layer 3	Bias Detection: Dynamic fairness audit for tracking bias disparities across demographic groups	Dynamic Fairness Audit Module	Input → Outcome data, group information → Output → Bias disparity measurements between groups
Layer 4	Rectification: Bias rectification loop adjusting model parameters to reduce fairness gaps	Bias Rectification Optimization Engine	Input → Bias audit results → Output → Adjusted model parameters for improved fairness
Layer 5	Evaluation: Continuous monitoring of fairness and performance based on stakeholder feedback	Continuous Fairness Evaluation Module	Input → Post-deployment feedback, performance data → Output → Ethical performance reports and adjustments

Pastor-Escuredo et al. (2022) argue that ethical frameworks should not only address individual-level biases but also consider systemic inequalities that exist within societies. This perspective highlights the need for AI systems to promote inclusivity and equity actively, particularly for marginalized communities that may be disproportionately affected by technological advancements.

9.4 The Role of Regulatory Frameworks and Governance

Effective governance and regulatory frameworks are vital for ensuring ethical AI development. These frameworks should provide guidelines for ethical practices while allowing for innovation and technological advancement. Wang (2023) suggests that ethical governance should be internalized within technology rather than imposed externally, creating a culture of ethical responsibility among AI developers and stakeholders.

The alignment of AI development with the United Nations Sustainable Development Goals (SDGs) provides a comprehensive framework for addressing social impact. By integrating ethical principles with the SDGs, AI technologies can be designed to contribute to global goals such as reducing inequalities, promoting sustainable cities, and fostering inclusive growth (Pastor-Escuredo et al. 2022). This alignment encourages developers to consider the broader societal implications of their technologies.

To formalize ethical considerations and embed accountability mechanisms within AI systems, we propose an Explainable Accountability-Aware AI Architecture, which combines SHAP-based interpretability, auditable optimization trajectories, and developer attribution mappings. This framework ensures that all decisions are not only explainable in real time but also retroactively traceable making it possible to identify how, when, and by whom biased behaviors emerged during the AI development cycle.

The system centers around the Explainability Attribution Function, defined using SHAP values (Shapley Additive Explanations):

$$\hat{y}_i = f(x_i) = \phi_0 + \sum_{j=1}^{n} \phi_j$$

where $\hat{y}_i$ is the model prediction for instance x_i, ϕ_0 is the base value (average prediction), and each ϕ_j represents the contribution of input feature j to the prediction. This decomposition provides full transparency over the internal logic behind AI decisions, allowing external auditors and users to inspect which variables drove critical outcomes (e.g., loan denial, risk classification).

To enforce accountability across the development team, we define a Traceable Parameter Optimization Path:

$$\theta^* = \theta^0 + \sum_{t=1}^{T} \Delta\theta_t \cdot \mathbb{I}_{d_t}$$

where θ^* is the final model parameter set, θ^0 is the initialization vector, $\Delta\theta_t$ is the parameter update at iteration t, and $\mathbb{I}_{d_t}$ is a developer assignment indicator tagging which individual or team d_t committed the update. This formulation enables post-training accountability assigning responsibility to model contributors at each stage of parameter evolution.

To ensure ethical principles are followed across iterations, we introduce the Ethical Compliance Constraint as a monitoring function:

$$\mathcal{C}_{\text{ethics}}(\theta_t) = \begin{cases} 1, & \text{if fairness and transparency metrics} \geq \text{threshold} \\ 0, & \text{otherwise} \end{cases}$$

This binary constraint ensures that any policy iteration violating fairness or explainability thresholds is automatically flagged for review and rollback. Transparency metrics may include entropy of SHAP distributions, while fairness thresholds may use SPD, DIR, or Equalized Odds metrics.

These components are integrated into a Responsible AI Reward Function in reinforcement-based systems:

$$R(s_t, a_t) = U_{\text{task}}(a_t) - \gamma_1 \cdot \mathcal{L}_{\text{bias}}(a_t) + \gamma_2 \cdot \mathcal{C}_{\text{ethics}}(\theta_t)$$

where $U_{\text{task}}(a_t)$ is the utility of action a_t, $\mathcal{L}_{\text{bias}}(a_t)$ measures outcome disparities, and $\mathcal{C}_{\text{ethics}}(\theta_t)$ serves as a compliance bonus. γ_1 and γ_2 are balancing factors that prioritize ethical behavior during AI learning and deployment.

This ethical accountability model transforms abstract values such as fairness, transparency, and responsibility into measurable components of the AI training lifecycle. It provides not only explainability for users and external regulators but also developer-level traceability, enabling a new standard of professional and institutional accountability.

Framework: Explainable Accountability-Aware AI Architecture (XAI-AAI)
Layered AI Architecture for Ethical Governance, Explainability, and Accountability in AI Systems

Layer	Main functionality	AI module(s) and components	Operational scope
Layer 1	Data Collection: Gathering stakeholder feedback and performance metrics for ethical evaluation	Ethical Data Preprocessing Engine	Input → Stakeholder feedback, performance data → Output → Preprocessed data for fairness and accountability evaluation
Layer 2	Learning: SHAP-based interpretability for model explainability and transparency	SHAP-based interpretability	Input → Model predictions and feature importance → Output → Transparent decision-making process
Layer 3	Optimization: Traceable parameter optimization path for accountability	Parameter Accountability Tracking Module	Input → Model parameters, developer assignments → Output → Traceable optimized model parameters
Layer 4	Compliance: Ethical compliance constraint to ensure fairness and transparency	Ethical Compliance Enforcement Engine	Input → Fairness and transparency metrics → Output → Flagged iterations for review and rollback if thresholds are violated

(continued)

(continued)

Layer	Main functionality	AI module(s) and components	Operational scope
Layer 5	Evaluation: Responsible AI reward function for balancing task performance, bias, and ethical compliance	Reinforcement learning	Input → Action decisions, fairness feedback → Output → Optimized decisions considering ethics, performance, and fairness
Layer 6	Feedback Loop for Ethical Refinement: Continuous monitoring, adjustment, and retraining to adapt to evolving standards	Continuous Ethical Adaptation Module	Input → Feedback data, performance evaluations → Output → Ongoing adjustments and retraining for continuous ethical alignment

Figure 9.1 presents frameworks for developing ethical AI that prioritize social impact, organized into four main areas: ethical frameworks, human-centered design, addressing bias and inequality, and regulatory governance. Ethical frameworks like Responsible Research and Innovation (RRI) focus on inclusivity and stakeholder engagement, while the AI Trust Framework and Maturity Model (AI-TMM) emphasizes trust, explainability, and privacy with structured metrics. Human-centered design involves community engagement to ensure alignment with public values and incorporates Explainable AI (XAI) to enhance transparency. Addressing bias and inequality requires using diverse datasets to mitigate bias and implementing ongoing audits for fairness. Regulatory and governance frameworks promote ethical responsibility within technology culture and align AI development with UN Sustainable Development Goals (SDGs) to maximize societal benefits. Each area is color-coded, linking subtopics to provide a comprehensive approach to ethical AI focused on social good.

The development of ethical frameworks for AI that prioritize social impact is essential in navigating the complex landscape of emerging technologies. By emphasizing human-centered design, stakeholder engagement, bias mitigation, and effective governance, these frameworks can help ensure that AI systems contribute positively to society. Ongoing research and collaboration among stakeholders will be vital in refining these frameworks and addressing the ethical challenges posed by AI.

9.5 Conclusion

This chapter has demonstrated that ethical AI cannot be relegated to abstract principles or after-the-fact audits; it must be engineered directly into the architecture of intelligent systems as a measurable, adaptive, and enforceable property. By weaving

Fig. 9.1 Frameworks for ethical AI development prioritizing social impact

together multi-stakeholder governance, trust-driven reinforcement learning, human-centered co-design, fairness monitoring, and accountability-aware explainability, we have shown that ethics can be formalized as both a technical constraint and a societal compass. The result is a transformative set of frameworks where inclusivity is not symbolic but computationally embedded, where fairness is not declared but continuously audited, and where accountability is not aspirational but traceable to every parameter update.

What emerges is a holistic vision of AI development that advances innovation without divorcing it from legitimacy, ensures transparency without sacrificing performance, and fosters equity without constraining progress. By aligning these models with global priorities such as the Sustainable Development Goals, the chapter makes clear that ethical AI is not a niche concern but a planetary imperative one that redefines the relationship between technology, governance, and society.

References

Wang, D. (2023). Towards Ethical Engineering: Artificial Intelligence as an Ethical Governance Tool for Emerging Technologies. *Computational Science and Mathematics Forum, 8*, 76. https://doi.org/10.3390/cmsf2023008076

Mylrea, M., & Robinson, N. (2023). Artificial Intelligence (AI) Trust Framework and Maturity Model: Applying an Entropy Lens to Improve Security, Privacy, and Ethical AI. *Entropy, 25*, 1429. https://doi.org/10.3390/e25101429

Pastor-Escuredo, D., Treleaven, P., & Vinuesa, R. (2022). An Ethical Framework for Artificial Intelligence and Sustainable Cities. *AI, 3*, 961–974. https://doi.org/10.3390/ai3040057

Sargiotis, D. (2024). Ethical AI in information technology: Navigating bias, privacy, transparency, and accountability. Advances in Machine Learning & Artificial Intelligence, 5(3), 01–14 https://doi.org/10.33140/AMLAI.05.03.03

Sargiotis, D. (2025). Fostering Ethical and Inclusive AI: A Human-Centric Paradigm for Social Impact. International Journal of Research Publication and Reviews, Vol 6, no 1, pp 3754–3779 https://doi.org/10.55248/gengpi.6.0125.0529

Conclusion

This book examines the multifaceted impact of Artificial Intelligence (AI) on urban development, focusing on promoting ethical, equitable, and socially responsible AI integration. This book explores how AI, when applied within public policy frameworks, holds the potential to increase efficiency, transparency, and fairness across various sectors, including healthcare, resource distribution, and public services. By leveraging AI in agenda-setting, policy formulation, and real-time service adjustments, cities can optimize public administration and enhance citizen engagement. However, the book cautions that AI deployment within public services introduces risks, such as job displacement due to automation and the digital divide, which require proactive reskilling strategies and policies that ensure equitable access to AI benefits.

A central theme of the book is the emphasis on incorporating ethical considerations into AI systems, with a focus on fairness, transparency, and accountability to mitigate potential algorithmic biases. Privacy concerns are particularly highlighted, as AI-driven surveillance technologies often collect vast amounts of personal data through urban networks, posing challenges around data rights, transparency, and citizen consent. The book argues that addressing these privacy concerns requires robust regulatory frameworks, grounded in principles that protect citizens' data and ensure clear governance of AI technologies within public domains. Furthermore, the risks of algorithmic bias and digital exclusion are underscored, as AI systems can inadvertently perpetuate or even magnify existing social inequalities if not designed and monitored with inclusivity in mind.

This book also discusses the transformative role of AI in delivering equitable and efficient public services. Automation and predictive analytics in public administration can streamline processes and enable more data-driven decision-making, enhancing accessibility and efficiency in service delivery. However, the ethical challenges associated with automated decision-making particularly in areas like healthcare and welfare allocation are profound, necessitating the development of governance frameworks that prioritize transparency, accountability, and citizen rights. The book advocates for inclusive policies that bridge the digital divide, ensuring that AI-driven public services are accessible to all societal groups, including marginalized communities who may otherwise be excluded from digital advancements.

D. Sargiotis, *Empathy in Algorithms*, https://doi.org/10.1007/978-3-032-32128-2

The book's examination of the digital divide reveals that socioeconomic and geographic disparities in technology access are key barriers to fully inclusive AI adoption. It discusses various strategies, such as public-private partnerships, community-based initiatives, and targeted infrastructure expansion, to address these disparities. These approaches not only provide affordable access to technology but also foster digital literacy through educational programs, empowering underserved populations to participate in the digital economy and benefit from smart city innovations. Ensuring inclusivity and equity in AI requires collaboration across government, industry, and communities, reinforcing the importance of a human-centered approach to urban AI integration.

In promoting a socially inclusive framework, the book underscores the importance of human-centered AI (HCAI) and equity-by-design principles that prioritize human values and needs throughout the AI lifecycle. This approach entails engaging diverse stakeholders citizens, ethicists, technologists, and policymakers in the development and deployment of AI systems to ensure that technological advances align with societal goals and values. The book highlights the potential of HCAI to bridge the gap between AI's technical capabilities and the social contexts in which it operates, enabling AI systems that not only enhance urban efficiency but also respect and uphold human rights, diversity, and inclusivity.

The book presents a comprehensive framework for integrating ethical AI in urban environments, advocating for transparency, accountability, and inclusive governance as essential pillars of AI-driven urban development. As AI becomes increasingly embedded in urban infrastructure and public administration, addressing the ethical, social, and policy implications is crucial for building cities that are not only technologically advanced but also just, accessible, and equitable. This book calls for a global commitment to responsible AI, with international cooperation and regulatory oversight to ensure that AI's benefits are distributed equitably, and its risks managed transparently, for a sustainable and inclusive urban future.

The manufacturer's authorised representative in the EU is Springer Nature Customer Service Centre GmbH, Europaplatz 3, 69115 Heidelberg, Germany. If you have any concerns regarding our products, please contact ProductSafety@springernature.com

Printed and bound by CPI Group (UK) Ltd, Croydon, CR0 4YY
08/10/2026
02229989-0003